ABOUT THE AUTHOR

György Moldova is the author of seventy-two books—nonfiction books on social themes, as well as novels, satires, story collections, and dramas—that have together sold more than twelve million copies over the decades, more than that of any other Hungarian writer. His achievements have been recognized with the Kossuth Prize, Hungary's most prestigious literary honor, and his work has been translated into several languages, including English, German, Russian, and Chinese. Born in 1934, he lives in Budapest.

ABOUT THE TRANSLATOR

David Robert Evans was born and raised in Oxford, England, and was educated at Worcester College, Oxford University. He is half-Hungarian and has devoted a significant part of his life to Budapest. His numerous translations from Hungarian to English include stage plays, screenplays, and other literary works, as well as a wide variety of nonfiction.

BALLPOINT

A Tale of Genius and Grit,
Perilous Times, and the Invention
that Changed the Way We Write

GYÖRGY MOLDOVA

Translated from the Hungarian by David Robert Evans

 New Europe Books

Published by New Europe Books, 2012
15 Doanes Lane
North Adams, Massachusetts 01247
www.NewEuropeBooks.com

First published, in Hungarian and English, in 2001 by ICO Inc., one of Central Europe's largest manufacturers of stationery and writing instruments, to mark its fiftieth anniversary.

ISBN: 978-0-9825781-1-7

Library of Congress Cataloging-in-Publication Data

Moldova, György, 1934-
[Végtelen vonal. English]
Ballpoint: A Tale of Genius and Grit, Perilous Times, and the Invention that Changed the Way We Write / György Moldova; translated from the Hungarian by David Robert Evans.
p. cm.
"First published, in Hungarian and English, in 2001 by ICO"—T.p. verso.
ISBN 978-0-9825781-1-7
1. Ball-point pens—Hungary—History. 2. Bíró, László József, 1899-1985. 3. Goy, Andor, 1896-1991. 4. Inventors—Hungary—Biography. I. Title. II. Title: Ball-point.
TS1267.M65 2012
681'.6—dc23
2012008355

Printed in the United States of America on acid-free paper.

1 3 5 7 9 10 8 6 4 2

BALLPOINT

A Tale of Genius and Grit,
Perilous Times, and the Invention
that Changed the Way We Write

In memory of László Bíró and Andor Goy

"The history of the ballpoint pen is a heroic one. It is a tale of technical skill, bravery, and unstoppable inventive ambition, of enterprise in the face of risk; it is a tale not without its fair share of greed, intrigue, and jealousy. Along the way we meet multimillionaires as well as unsatisfied ambitions and ruined lives. Serious businessmen and talented scientists have as much a role in the turn of events as do tricksters and simple mercenaries."
—*Fachzeitung Papier*, Baden-Baden

PREFACE

Neither of the two protagonists of our story are alive today. László Bíró died in 1985 in Buenos Aires, Andor Goy in 1991 in Budapest. Although we have no recourse to their living testimony, they both, fortunately, left biographies for posterity. Bíró dictated his memoir into a tape recorder; this was edited into a successful book by the Argentine journalist Hector Zimmerman titled *Una Revolución Silenciosa* (*Silent Revolution*) and published in Hungarian translation in Budapest in 1975. Goy's memoir, on the other hand, entitled *The Real Story*, is fading away, unpublished to this day. In these very personal recollections it is not hard to discover a strong element of self-justification, and a certain distance from hard fact. Bíró is open in admitting, "The readers of this book should never forget that what they hold in their hands is a hopelessly biased work. What I recount is the truth, but it is probable that the facts and persons presented are not in reality quite as I describe them." As for Goy, it was only in his later years that he began to put his memories to paper. His intellectual powers had of course deteriorated over the years, as evidenced by occasional factual errors, especially with regard to dates. To mention just two of his gaffes, he refers to the state of Israel with reference to an event in 1942, when Israel was in fact only founded in 1948; elsewhere he confuses two Argentinean presidents, De Justo and Peron. Yet despite these qualms, in my journey through this whirlwind of events, it is these two manuscripts that have been my Ariadne's thread.

The two men describe events that only they lived through, and only they know—and often I relay these verbatim. While using their texts, I have made every effort to be faithful to their personal tone and to the way they relate the various episodes in the form of dialogue scenes. In addition to their memoirs, I have made use of a number of other sources: newspaper articles, legal records, contracts, and other works relating to the subject. My primary evidence has been in the form of conversations with patent agents, historians of science, Hungarian and Argentinean diplomats, and those involved in the production of ballpoint pens. The greatest help, of course, was that which I received from the relatives of the book's protagonists. Goy's daughters Gabriella and Krisztina made me welcome in their father's old house; Mariann Bíró allowed me to pay her a visit in Buenos Aires.

Even in command of all this information, the story demanded continual interpretation. It was necessary to define its place on the map of Hungarian history of the twentieth century, and to square often contradictory claims about what occurred—without taking on the role of judge and jury.

I was not in a position to compromise my objective of turning this tale into a readable book. This explains, I hope, why the entire text is colored by my assumptions and by the workings of my imagination. I would ask the reader to consider it both a historical manuscript and a novel.

If it fails to make the grade in either respect, or indeed in both, I can only quote the great Russian mathematician Lobachevsky when he says, "The real problem with books about geniuses is that they are not written by geniuses."

CHAPTER I

The first character to step onto the stage in our story is the younger of our heroes, László Bíró. Before he does so, however, it would be worth taking a glimpse at the backdrop from which he emerges (Bíró was born in 1889), that of Hungarian intellectual life at the turn of the twentieth century.

By the entrance to 68 Fő Street in Budapest, home to the Union of Technical and Scientific Associations, there are two plaques. One lists Nobel laureates of Hungarian origin, together with the year in which they accepted their prizes:

Philipp Lenard—1905
Robert Bárány—1914
Richárd Zsigmondy—1925
Albert Szent-Györgyi—1937
George de Hevesy—1943
Georg von Békésy—1961
Eugene Wigner—1963
Dennis Gabor—1971
John C. Polanyi—1986
George A. Olah—1994
John C. Harsanyi—1994

The other plaque, entitled "Hungary's Greats," lists names like John von Neumann, Leó Szilárd, and Theodore von Kármán. All of these figures developed or laid the foundation for their scientific work in the first decades of the twentieth century.

To the outsider, it may seem hardly credible that a country of ten million people could have given the world such an array of geniuses almost at once. Many attempts have been made to provide an answer to this mystery. Some look for explanations on nationalist grounds—in the special, possibly even genetic, talents of the Hungarian people. Unfortunately for the exponents of such views, nationalists in other countries make similar claims, and the world is not very interested in judging between them. Further, they are incapable of giving a suitable explanation for why this immortal national genius does not display itself on a continuous basis, irrespective of the twists and turns of history. It seems we should heed the words of Ferenc Nagy, the biographer of Hungary's Nobel Prize winners, when he suggests that this scientific sensation was a result not of biological hardware but of sociocultural software. One of those concerned, Georg von Békésy, makes the following observations:

> Everyone poses the question: how can it be possible that a country as small as Hungary is home to such a large number of scholars of international renown? Many Hungarians have offered solutions to this puzzle; I myself am not sure what the exact answer is. The one thing I can say is that the years of my life I spent in Switzerland were so calm and settled that I felt no need to fight in order to survive. The situation was quite different in Hungary, where life was a continual struggle for just about everything, though this struggle was such that no one actually perished in it. One could survive, one sometimes won, sometimes lost, but it was never decided completely and finally, at least not in my case. There is a need for such struggles and throughout its history Hungary has had its fair share of them. It is most likely a Hungarian trait that has developed over the generations that if they have to die, they should at least lay down their lives for king and country, and in a number of cases it is this that keeps them alive.

After Békésy's elegant but rather vague words we should attempt to get a little closer to the essence of this phenomenon. It is most likely that the key to the puzzle lies in the unusually high standards of education for children of the Hungarian elite at the turn of the twentieth century.

Norman Macrea, a contemporary editor of the *Economist*, described the circumstances of the period as follows: "Budapest was the fastest developing metropolis in Europe. This city produced a legion of scientists, artists, and would-be millionaires, in numbers only comparable with the Renaissance city-states of Italy. The elite secondary schools of Pest were without doubt the most successful schools history has ever seen."

The foremost among such elite schools was the Lutheran Lyceum, alma mater of Leó Szilárd, later one of the pioneers of the atom bomb; Eugene Wigner, the Nobel Prize–winning physicist; John von Neumann, father of the modern computer; and a number of other world-famous scientists.

Despite its name, the students at the Lutheran Lyceum were by no means limited to one creed: it was the wealthy Jewish community, which played such an important role in the economic life of the period, that was most prominent in sending its children there. Of the 665 students enrolled in the academic year 1915–16, 352 were of Jewish origin; and in 1917–18, out of 669 students, this figure was 373.

The parents of these children did not wish to send their sons to Jewish educational institutions, full to the brim as they were with religious commitments; they gave greater priority to the study of modern technical subjects than to the Torah and the Talmud, considering science, not religion, to be the future. They also sought to preserve their close-knit family lives; while they were in a position to afford education for their children in England or Germany, they were reluctant to send them abroad. And so they were happy to make a substantial sacrifice to be able to secure high-quality education for their children in Budapest. The fees at the Lutheran Lyceum, for example, exceeded the salary of a low-ranking civil servant.

Such a financial foundation made it possible for the school to attract the very best teachers available. Such were the stipends they offered to a number of academics, like mathematician László Rácz and physicist Sándor Mikola, that these people opted to teach at the school rather than lecture at a university.

This emphasis on the "real sciences," as they were called at the time, naturally had its desired effect—with half of those leaving the school choosing to follow careers in technical or scientific fields, one in five opting for medicine, and only every one in twenty pursuing further studies in the humanities. The school's leadership was well aware, however, that the teaching of the scientists of the future would not be complete if it were limited to technical and scientific subjects; students had also to be versed in history, law, and economics. This approach was consistently and successfully used in teaching, producing broad-minded and versatile alumni. When Eugene Wigner collected his Nobel Prize, his first response was to give his thanks to the Lutheran Lyceum, and to his teacher László Rácz in particular.

There was only one field from which this brilliant young group kept its distance: that of politics. Although members of the Jewish bourgeoisie played a key role in the movement that led to establishment of the Hungarian Soviet Republic—we need only mention the names of Béla Balázs or György Lukács—these young Turks refrained from getting involved in revolutionary movements, regarding them as beneath their high intellectual standards.

The right-wing government that came to power after the war did little to reward the distance this group had kept from politics. As early as 1920 it instituted the first law of numerus clausus, under which: "From the academic year 1920–21 only those individuals can apply to universities of sciences, the Technical University, Budapest University of Economics, and to the Academy of Law, who are completely reputable in respect of their national allegiance and morality. . . . Care should be taken that the proportion of students belonging to one or other of the various ethnic groups resident in this country should if possible reflect the proportion of that group in the country as a whole."

For most young people of Jewish origin, this legislation closed the doors on a potential career in Hungary. Universities saw the burgeoning terror of groups with fascist sympathies, with physical assaults taking place even in lecture halls. Those able to do so left the country.

Obstructions facing young students of science were not limited to their university careers, with the regime continuing to discriminate against them after their graduation. John von Neumann was never allowed to become a member of the Hungarian Academy of Sciences; the Academy decided to give its highest prize not to Albert von Szent-Györgyi, who discovered Vitamin C, but to the ethnic theories of Ferenc Orsos, member of the fascist Arrow Cross movement. As one contemporary notes, "The torture of the talented was to continue after 1918. By the time Hungarian physicists were in a position to make their mark on physics, they were all abroad."

This was not just the case in physics. Albeit at different times, all the members of this unique generation of Hungarian geniuses made their departure not only from Hungary, but from the Old World altogether. The "Big Six" assembled in the United States: John von Neumann, Edward Teller, Theodore von Kármán, Leó Szilárd, Eugene Wigner, and Dennis Gabor all set off for international fame, some in the direction of the Nobel Prize. Dennis Gabor was justified in observing in a letter, "Hitler did not do me a complete disservice in chasing me out of Europe."

While this statement is certainly emotive, it cannot fully divert our attention from the sad fact that not only did these scientists lose their homeland, but their homeland also lost a set of irreplaceable intellectual treasures.

CHAPTER 2

I have no intention of mentioning the ballpoint pen and the computer or the hologram in the same breath. Neither do I want to consider László Bíró as being on a par with John von Neumann or Dennis Gabor, whose names are most closely associated with the latter milestones in technological achievement.

But what would Bíró have been like in the flesh? One surviving photograph depicts a short, prematurely balding man, with a dome-shaped forehead and protruding chin, holding spectacles in his right hand. Francis Loring Sweet, who would later become his son-in-law, wrote that "he had extraordinary blue eyes; they were emotive, but not prying, and free from any kind of aggression. He concentrated all his attention on hearing and listening, on thought and the exchange of thoughts with others. He had the internal space and the patience to put his finger on the essence of things."

In the course of my research, I had cause to speak with a number of people who knew László Bíró personally: they described him as a radiant, dynamic personality. János B., former Hungarian ambassador to Brazil, related a tale from their friendship. B. had for a long time tried to master the science of driving an automobile, with no success. He had, in fact, almost given up hope of ever succeeding. On the occasion of a trip made jointly with Bíró, he explained his inhibitions in this regard. Bíró immediately handed him the wheel. The effect was like a cure on B.: he took control, and, surprising himself more than anyone else, managed to drive with perfect competence. Another Hungarian diplomat was of the view that Bíró was the country's greatest genius of the last two hundred years, as inventor, painter and thinker.

Such opinions seem exaggerated—it is no coincidence that they were voiced by diplomats—but there is no doubt that in Bíró we are faced with a creative, fiercely independent character, who bore the "mark of genius."

Bíró was born into a Jewish petit bourgeois family in Budapest. His father, Mátyás Bíró, an accredited dental practitioner, had a successful surgical practice in the Lipótváros district of central Budapest. His talents exceeded the tight confines of his profession, and, as a hobby, he produced a few inventions, like the "rubber porter," which allowed the front door to an apartment to be opened from any of the rooms. As if Fortune wished to give advance notice of the idea behind László Bíró's most significant endeavor, the elder Bíró experimented with a new kind of pen: instead of filling the cartridge with ink, he filled it with water, which, flowing through a thin tube, dissolved an ink cartridge core. This new device was not a success, however, because it proved incapable of providing a continuous and even flow of ink.

The family's first child, György, was born in 1897. In Bíró's opinion he was a hard-working model child who, when preparing for exams, would walk around the kitchen table a thousand times with a book in his hands. Yet he did not pursue his way in life with any particular imagination. He fulfilled his family's wishes, climbing one rung above his father in the social ladder, gaining a university degree in dentistry.

The first hero of our story was born in 1899. From early childhood, László Bíró was the family's *enfant terrible*; he would fight, hang out in the street, and find all kinds of ways to skip classes at school. At the age of eighteen, he was called upon to perform military service, but never made it to the front, as the Austro-Hungarian war machine was falling to pieces in the meantime. Unlike his older brother, László was not too bothered about methodically building a livelihood for the future. He did complete a few semesters at the University of Medicine, but the only aspect of the medical profession that captured his imagination was hypnosis.

He explains his failure to finish his course: "I was the first person in Hungary to deal seriously with practical hypnosis. I made so much money out of it that I lost all interest in continuing my medical studies."

This chopping and changing of professions was to continue in his later years: "I have been a medical student, a race-car driver, a graphologist, a biological researcher, and an insurance agent, but I have also been a trucker, a painter, a journalist, a book publisher, a sculptor, and an inventor. It is fair to say that my profession is not having a profession." In our cynical times we would find it difficult to believe that there still exist such Renaissance men in the modern world and would be a bit suspicious of such claims as reeking of self-promotion. Looking a little closer at Bíró's words, such suspicions would not be so unreasonable: his skills as a race car driver refer to a single occasion and his biological research to the dissection of a couple of frogs. But there is no doubt that as a painter he was a member of the most select group of artists. It is not my place to begin an analysis of Bíró's whole career, still less to provide an assessment of it. I wish to concentrate on just one of its many strands: his activities as an inventor.

His first invention, the "water fountain-pen," was patented in 1928; we know nothing else about it, but can hazard a guess that it was a further development of his father's earlier idea.

Bíró's next invention dates back to 1930. Called the Mesemosó ("Fairytale Washer"), it was, in his words, "a perfect home washing-machine, the energy needs of which could be supplied by putting it on an ordinary kitchen stove. There was an automatic apparatus to signal when the whiter than white clothes were ready to be rinsed." The accompanying brochure enticed potential customers with the promise that the costs of using the machine were but a tenth of those incurred by washing by hand; nevertheless the machine failed to prove particularly successful and rapidly disappeared from the shelves.

Neither did Bíró's following invention, the electromagnetic mail dispatch system, bring him much more in the way of success. As its instructions described, "The dispatch tube systems in use today were built on the principle of the suction properties of a vacuum, but we see no reason why this system of evacuated tubes should not be replaced with a postal apparatus using the infinitely faster and safer method of electromagnetism, an apparatus that can be of any length."

It is clear from the wording of the details concerning the system's operation how well Bíró, at the time employed in journalism, was aware of, and capable of thinking through, the most practical, technical solutions. Experts consider this invention to have been ahead of its time: the essence of the idea was later adapted by Japanese railways in the conveyance of trains. In short, it was fortunate that Bíró did not rely on income from inventions for his livelihood.

Menyhért Kraszner, owner of the Japan Café that Bíró frequented on Andrássy Boulevard, once announced a competition to his guests, asking them to summarize their curriculum vitae in as few words as possible. Bíró won the competition with the following entry: "When he was good he was good, and when he was bad he was bad." He was enjoying a "When he was good he was good" stretch of his life at the time, making a lot of money out of various customs ventures and publishing numerous articles in newspapers.

Bíró devoted much of his time to visiting clubs with his friends. In the course of the eventful frolics of one particular evening, Bíró bought a red, twelve-cylinder Bugatti automobile from a well-known actor. At the moment the deal was struck, and noticeably under the influence of champagne, he also offered to use the car to enter a race on Sváb Hill in Budapest some two weeks later. Bíró had the slight handicap of never having sat behind the wheel of a car in his life and was not even acquainted with the most basic aspects of driving.

His pride prevented him from shirking the challenge, however, and he took lessons from an old chauffeur friend. It took Bíró a while to master the necessary moves, and he would often attempt to change gears without putting his foot on the clutch, resulting in a terrible noise like that of a nutcracker. His teacher would always punish his mistake by stomping on his foot; every evening Bíró would have to go home and bandage it up. Bíró won the race, or so he claims, but was never able to get over the fact that a driver has to put his foot on the clutch every time he changes gear. And this was how, after much deliberation, he hit upon the idea of the automatic gearbox.

Bíró soon outlined the designs for his gearbox. He showed them to an engineer by the name of András Buday, who was impressed and was willing, in return for half the proceeds, to finance the development of the idea. It took over a year to construct a satisfactory model, which was designed to fit a motorcycle with a sidecar.

The partners set out to sell their invention. They showed it to the technical director of the Ganz factory, who thought highly of the design. Though he had no wish to deal with it himself, he was happy to write Bíró a letter of reference addressed to the General Motors plant in Berlin. A contact was successfully drawn up and GM designated Germany's Avus highway as the location for a test run. Bíró had no faith in the intentions of this industrial giant, however, and noted in his book that he tried to prepare for every possible objection in advance: "I called the Hungarian Automobile Club to get them to seal the gearbox housing. This is how I wished to prove that the part had not needed any kind of maintenance throughout the whole 600-mile journey, along a route often marked by high mountain regions. When they finished, I climbed on the 350cc motorcycle and, with a 200-pound engineer in the side-car, set off on the road to Berlin."

Four test drives were held at the ring at Avus, and although the gearbox worked perfectly on every occasion, the German engineers wished to examine its construction at close hand. Bíró was not willing to satisfy this request, as he had only submitted a patent application in Budapest and was worried that they would steal his idea. He argued that there was no need for them to inspect the gearbox, which in its sealed housing could not possibly have been tampered with.

Leaving the matter in the air, Bíró returned to Budapest on his motorcycle. It was there that he received a contract proposal from General Motors: they offered him half a percent of the sale price of units produced, and an advance of 200 dollars a month over a period of five years—half of which would be for Bíró, half for his partner Buday.

Even 100 dollars was a lot of money in 1932; converted into the Hungarian currency of the time, it was worth 500 pengős. As a guide, a daily wage in the agricultural sector would have been about one or two pengős; to give a more general picture, there was a popular song entitled "Two Hundred Pengős Coming In and You'll Never Look too Thin."

Bíró, who had married in the meantime, was truly captivated by such an offer and the trouble-free family lifestyle that it would guarantee. This explains the fact that, with his partner, he signed the agreement without showing its details to a lawyer or patent agent. Their inexperienced eyes failed to notice the two legal traps in the contract: it neither made mention of the date on which production would commence nor did it mention the number of gearboxes to be produced. By the time they came to their senses, and sent an urgent letter to General Motors, the latter was well within its legal right in citing the text of the contract. *Pacta sunt servanda*—the conditions must be met.

"Only later did we learn," Bíró relates, "that the reason behind General Motors' action was that if our invention had been put to use, huge numbers of components produced years in advance would have become unsellable. In addition, the factory had been experimenting with a hydraulic automatic gearbox for years, for which it had higher hopes than for my mechanically-based device. Their only reason for signing the contract with us was to stop the competition from snatching up our invention."

Across the world occurrences like this were not so rare. The cartel of German bottle producers purchased Owers' bottle production machine without using it for a number of years, lest the new device, capable as it was of producing at lower cost, should cause a fall in prices and hence a drop in profits. The same story could be seen in the case of the long-life light bulb and the American light bulb cartel.

This was, however, the first time that Bíró had come face-to-face with the "big sharks" of the patent world, who by devious means had shortchanged him of the license payments he considered his due. It seems reasonable to think that he must have made up his mind never again to fall victim to them.

Bíró set to work on a new task. He began planning a high-powered turbo engine, but his partner lost his entire estate in a family catastrophe, and they had to give up the experiments for lack of funds.

CHAPTER 3

E ven amid such active technical experimentation, Bíró did not give up his journalistic work. He was editor of the fine art journal *Hongrie–Magyarország–Hungary* until it was discontinued, when he moved to the opposition weekly *Előre* (Forward).

He was only rarely to be found in the office; Bíró wrote his articles in the Japan Café. The Japan, named after its ornamental Oriental wall tiles, was but one of a number of similar establishments. As early as the turn of the twentieth century, Budapest could rightly think of itself as a "café city"; there was hardly a street without a café or coffee shop. Even the smallest of social groups had their own regular haunts: market women, beef traders, assistant barbers, as well as blind musicians or pickpockets.

As someone once wrote, the coffeehouse was life's greatest ground floor and market place—half street, half shelter. A more cynical observer once added, "Coffeehouses have two advantages. You are not at home, yet neither are you out in the fresh air."

The coffeehouses would often keep their doors open day and night; many were famous for having thrown their keys into the Danube on their opening night, signaling that they would never close. They strove to provide as wide a selection of services as possible; there was a time when the famous New York Café was home to both a hairdresser and a pawn-shop. Café owners would provide their own encyclopedias for their guests to use; on the shelves of the foyer one could also find a full list of Budapest's residents and their addresses, a handbook of financial services, and the train timetable.

Some regular customers would even have their mail delivered to their café. An old actor by the name of Kálmán Csetery sat at the same table for twenty-five years; after his death his gravestone was carved out of the table's marble surface. It was also possible to view the developments of history from the comfort of one of the café chairs. From time to time history would come to you: one of the best examples of this was when the revolutionary youth flew their flag in the Pilvax café in 1848 amid Hungary's ill-fated uprising against the Habsburg Empire.

During World War I, families would say goodbye to their sons at one of the so-called "cafés of death" near the Keleti and Nyugati train stations—the Terminus, the Ilkovits, and the Wágner. To help the war effort the coffeehouses were obliged to surrender most of their metal cutlery. Names that referred to enemy countries were changed; it was in this way that the London Café came to be called the Berlin. Such measures would later become more extreme. It cannot be a coincidence that every regime of terror that appeared in the twentieth century began its work by ordering the closure of the cafés.

To return to the Japan Café of the 1930s: it was first and foremost a haunt for writers and journalists. There were many hundreds of publications for visitors to read, both domestic and foreign. Ink was kept in the cellar by the demijohn, together with proof slips, pens, and hand-blotters for the benefit of the guests. Bíró made use of all of these on a daily basis.

The question inevitably arises: what caliber journalist could Bíró have claimed to be? Paging through his articles to be found in old copies of *Előre*, we must admit that he was not one of Hungary's greatest wordsmiths. His works were conventional, even routine, with typically wordy titles like "Work in the bakery industry must be regulated and bread products sold by the kilo!," "Assistant railway officer inherits a hundred thousand forints," or "Unraveling the mystery of the fratricide and double murder in Aszód." It was in mini-feature articles that Bíró's brilliant mind could be recognized. His stories were largely fictitious: "Currency was smuggled inside nuts, but the customs guards broke open their shells to find hundred-pengő notes inside." "The British air force will only accept candidates who can hold their breath for at least 68 seconds." "Following a fight a barber carried a pocket-knife stuck in his head for two years without noticing it was there." The payments he received for his journalistic work were in line with its quality: While the esteemed writer Sándor Márai was paid 2,500 pengős a month at the Pesti Hírlap, Bíró was lucky if he made 200.

Presumably, Bíró himself had few illusions about his journalistic talents, though he never admitted this openly for any kind of self-deprecation was against his character. This did not mean that the laws of psychology did not apply to him. He held the great personalities in the profession in the highest esteem, like his childhood friend and later Hollywood scriptwriter László Fodor; and, in particular, László Lakatos, lead writer for the *Est* titles, whom Bíró referred to as "Master."

Lakatos's erudition was European in nature but personal in form. In a style akin to the famous Hungarian writer Endre Ady, his literary biography claims that his father was Voltaire, his mother Heinrich Heine, his favorite mathematician Beethoven, his favorite musician Einstein. He observed the strengthening of fascism in the Hungary of the 1930s with a clear mind.

Though Lakatos would spend all day slaving at his desk, in the short breaks he took he was only too happy to welcome Bíró to his table. He was no bigoted teetotaler, however, and on such occasions he allowed himself a glass of some noble spirit. In his view one had only to be careful that one never became so poor, stupid, or empty inside that one had nothing to live for but alcohol.

"It's time to get out of here, Edison," he would say to Bíró, "preferably while we can still leave with a full suitcase and not empty-handed. It's just finding the right moment that's the difficult part. As the South African saying goes: When the Jews leave, it is still too early; by the time the Portuguese leave, it is already too late. All in all, though, I think I'd rather make the first mistake than the second one."

While Lakatos was only seven years older than Bíró, he was such a respected figure in the young inventor's eyes that Bíró had difficulty bringing himself to question what he said:

"But Master, this is where I would like to spend my life."

"So would I, if only they'd let me."

One day Bíró visited the Japan Café, but searched for László Lakatos in vain; Lakatos had taken his own advice and emigrated to his much-loved France. Soon afterward László Fodor also said good-bye to his old friend: Bíró's situation began to look increasingly lonely. Although for the time being he was not in direct danger, he could not avoid contemplating his fate on regular occasion. A seemingly irrelevant event sped up his decision. As he wrote in his memoir:

> I would always buy cigarettes in the shop of an old widow whose husband had been honored with a number of medals for his bravery in World War I. This was why the authorities had awarded her the license for the shop. One day I learned that this license had been revoked when a member of her family married a Jewish girl. I thought that as a member of the editorial board of a Christian-minded weekly with the prestige of *Előre*, I could do something to help the poor woman.

> I paid a visit to an official at the relevant department in City Hall, who just so happened to be an old friend of mine. In the friendliest way possible, I recounted the story of the old widow and asked him if he could possibly intervene. Without batting an eyelid the whole time I spoke, he replied forcefully: "I would ask you to refrain from getting involved in such matters! Whatever happened to her is entirely her own business." Asking him to remember the humanitarian considerations of the case, I began to talk about the woman's age and the terrible difficulty of her situation, only for the official to interrupt me: "You know very well, Mr. Bíró, that you yourself are one of the 'undesirable elements.'" He added, "You would do better not to interfere in such matters!" I accepted what he had said without a word.

Yet, despite his silence at the time, it transpires from a statement Bíró made in 1980 that the scar of the official's insult remained with him for the rest of his life.

This was, as they say, the last straw. Bíró resolved to emigrate. He would not have been his confident self were he for a moment to have worried about whether he would find his feet in a foreign land. He did, however, need a certain amount of capital before he could successfully begin his life again abroad. His monthly hundred dollars from General Motors would not get him anywhere and he had no other financial resources to speak of. Only the proceeds from a new and successful invention could save the day, but how would such a result emerge from the pressure of his desperation?

CHAPTER 4

In his notes we find a number of explanations for how Bíró came upon the idea for the ballpoint pen. The idea was born, some sources would have it, when a couple of little balls on Bíró's desk rolled through some spilled ink and drew lines in front of his eyes. This account does not quite explain what these small balls were and how they came to be there on the table.

According to a second, no more credible, variant of the story, Bíró was sitting on the terrace of a café, watching children playing marbles on the pavement. He noticed that, after passing through a puddle, the marbles would leave a wet trail behind as they rolled further along.

Bíró himself remembers the inspiration for the discovery as follows. As the on-duty editor, he would often go to the press to check proofs. Although he used an expensive Pelikan pen, he regularly found himself in an awkward position: while he was working, the ink would all too often run out or dry up, blocking the flow to the nib. If he pressed the nib harder, it would split in two and his writing would become thick and ugly. As he disassembled and mended the pen by hand, he would jealously watch the rotary machines as they grumbled away: the ink would be spread evenly over the letters and dry the moment they touched the paper, leaving the newspapers to drop onto the table without a blotch.

Albert von Szent-Györgyi once said that to be an inventor one has only to see what everyone else sees and think what no one else thinks. Bíró was not satisfied with standing there looking jealous, of course; he could not stop thinking about how the model of the rotary could be used to design a replacement for the fountain pen. He would have been in a position to reduce the number of components and miniaturize them as necessary, but he could not overcome a more fundamental stumbling block: the cylinders were only capable of moving in one direction. We read in his memoir: "'What can I use instead of a cylinder?' I asked myself while sitting at my usual table in the café. The solution flashed across my brain like a bolt of lightning: a ball!"

He was so overcome by inspiration that he ordered his next coffee not with a splash of milk but with a splash of ball.

"What I was looking for was something that was entirely new among writing instruments," he writes elsewhere in his memoirs. "I knew the origins and development of writing tools in considerable detail, from symbols inscribed in stone, through the brushes used by the Chinese to the split goose quill, from nibs made of steel, gold, and iridium to the fountain pen. But none of these precedents provided me with any kind of guidance for my work."

Here, thanks either to ignorance or to an exaggerated sense of self-importance, Bíró forgets to mention that his idea was neither exclusive nor new; it did, indeed, already have quite a history. An American technician applied for a patent for a writing instrument based on a similar idea as early as October 30, 1888. He was followed by an impressive list of inventors: the German Anton J. Sheaffer registered a patent for a ballpoint pen in October 1901, followed by Michael Braun in October 1911. At the Dresden trade fair of 1924, G. L. Lorenz presented a finished ballpoint pen with the brand name Mungo. There are precedents to be found even in Hungary. Dr. Dezső Ránki and railway engineer Ödön Hajdú patented their version of the idea in 1934. It is hard to believe that Bíró was unaware of any of these.

However strongly we might be tempted to accuse Bíró of plagiarism, the charge does not hold up; neither does that of patent fraud. At this point in the story, in 1936, there did not exist a ballpoint pen that worked properly: one type would leak, while another would cut out during writing or display other problems. It would be another decade before the first perfectly working specimen would appear, bearing Bíró's name. However it might vary in other ways, patent law in different countries unanimously states that there is no legal basis to be founded on partial results: it is the person who makes the final step who is recognized as the inventor. Full acknowledgment, in other words, goes to the one who "puts the dot on the i." In the eyes of the American system, moreover, it is of no significance who had the idea first. A good example of this is provided by those who discovered the theory behind the electric light bulb, but lacked the resources to put it into practice. Edison, on the other hand, only took up the challenge at a later stage, but was equipped with the apparatus necessary for its implementation, and so it was he who successfully registered the patent.

Bíró swallowed himself up in this new enterprise with all the energy he could muster, undeterred by his total lack of experience in the area. To conduct the necessary experiments he had to put together his team. The first person he recruited was his brother György, who as a dentist had some experience in chemistry, and was thus entrusted with the task of finding suitable material for the dye. This proved a harder nut to crack than they thought, even though they sought the opinion of a university professor of applied chemistry.

"Professor," began György, "we are looking for a dye that remains fluid in the cartridge but dries as soon as it touches the paper."

The esteemed academic could not hide his indignation.

"My good sir, there are two kinds of dye: those that dry quickly, and those that dry slowly. What in heaven do you mean by a dye that makes up its own mind when it should dry fast and when it shouldn't? This is *contradictio in adiecto*, contradiction in the premise itself—like ice that is hot and engraveable. It does not, and cannot, exist."

Undeterred by this outburst, the elder Bíró swallowed himself up in the literature, occasionally bobbing up to make profound announcements about surface tension, viscosity, and degrees of saturation.

During the whole six years of experiments that were eventually required to bring success, it was consistently the search for a suitable dye that proved the greatest obstacle. It was at least possible to clarify the fundamental principle that the pen would work not with ink but with a specialized paste, but dyes otherwise used for printing or in inkpads needed further and further alterations in order to be used in the tests.

"If on one day the liquid appeared too thick, by the next day it would be too thin, and after a while it would become thick again. It was only much later that we became aware that the changes displayed by the dye material depended on its temperature. How can the design of the ballpoint pen be constructed around a dye material that seemed to change on a continuous basis? Moreover, the aniline would often crystallize, rendering the entire mixture unusable."

These were by no means the only difficulties Bíró faced. This period was more characteristic of the second part of his earlier prize-winning curriculum vitae, "when he was bad he was bad": he was running out of money and beginning to have trouble financing even his most urgent expenses.

Bíró claims in his autobiography that he registered a patent for his original plans for the ballpoint pen, thus guaranteeing priority for his work over anyone else's for at least a year. It makes little difference to the development of the story, but we should note here that there is no mention to be found of this registration either in the official record of patents or in any other sources.

Bíró could not have afforded even a fraction of the cost of the full process of patent registration. Those who phrased their list of requirements in their own words could easily be outdone by competitors exploiting any weak points. For example, if an inventor were to patent a medicine composed of 80% material A and 20% material B, all a competitor had to do was to change the proportion to 81% and 19% in order to escape the jurisdiction of the patent. In order for the wording not to be vulnerable in such a way, it was necessary to hire a professional patent agent; this would cost two to three hundred pengős. This was on top of the twenty-pengő registration fee and twenty-pengő tax, to mention only the smaller expenses involved.

There was also the significant cost of maintaining an existing patent—a cost that increased constantly, from twenty-five pengő in the first year to two thousand pengő in the twentieth and final year. Delayed payment would incur a 100% surcharge; this was all only the cost of a domestic patent.

Bíró was not in a position to continue without access to outside financial resources. He added a third person to his team, his childhood friend Imre Gellért, who owned a travel agency on Nádor Square. Gellért was in a position to give a certain amount of help both financially and in solving some technical problems.

Bíró continued to control the work on the project, devoting every minute he could spare to his efforts. He hung a large sketch of the imagined pen above his desk. Staring at this drawing, he would think and rethink every possible way in which the pen could work, the problems that might be encountered, and ways in which they might be avoided. The possibilities were endless.

Bíró would sometimes construct a prototype, but these were always far from being perfect. The pen had to be held in a precisely vertical position, and even so, the movement of the ball was not smooth; at most it could draw a few hesitant circles. If he tried to put greater pressure on the thick liquid with a piston, the ball would shoot out of its position like a bullet, the dye spraying in every direction. After a while the walls of the small laboratory were covered with the dye, as were the clothes, faces, and hands of those carrying out the experiments.

Bíró's financial woes grew and grew. The storm clouds of war were not good for the tourism industry; business at Gellért's travel agency was slack, and he had less money to invest in the research. In 1937, Bíró's contract with General Motors expired, and so did the hundred dollars they sent him every month.

The only possible way to win further sponsors to the cause was to produce a working model of the pen that could be used for demonstration purposes. Bíró made every effort to keep costs to a minimum, but there were some vital components that he simply had to order from Switzerland. With these, and some components of his own, a Budapest precision instrument maker produced a few specimens.

CHAPTER 5

The year 1938 heard the whole of continental Europe shudder to the rhythm of German soldiers' stamping boots and victorious marching songs. Hitler's troops first entered Austria after the Anschluss, and after signing the treaty of Munich they marched into the Sudetenland, formerly a part of Czechoslovakia. Not satisfied with these conquests, they prepared for further aggression.

The impact of these events in Hungary was not difficult to perceive. The authorities were of the belief that an alliance with Germany would help them realize their revisionist goals of reclaiming the territories Hungary had lost under the Treaty of Trianon following World War I. As preparations for war began, extreme right-wing movements won greater and greater support: the Arrow Cross Party enjoyed 20% of the popular vote in the parliamentary election.

In Gorky's words, "Antisemitism is the Esperanto of fascism; it is the language by which it spreads around and infects the world." In the light of the so-called Nuremberg laws of 1935 in Germany, similar legislation against Jews was also passed in Hungary. The first Jewish law was passed by parliament on May 29, 1938; the second scarcely a year later. These measures limited the proportion of Jews in economic and cultural life, especially the media, and excluded them from work in the public sector entirely.

Bíró was unfortunate enough to experience these changes first hand. His articles could no longer appear in *Előre*. Furthermore, he had to face the fact that time was not on his side. Most of all, he feared that Hungary's borders would close and that he would be unable to make arrangements for his family to emigrate as he wished. His attempts to find out which countries might accept them were in vain; foreign embassies began to close their doors to applicants, only awarding visas in very limited numbers.

On one occasion—quite possibly after another fruitless round of inquiries—he made a visit to his brother's dental practice. György Bíró happened to be tending to a strikingly elegant and attractive woman named Mária Pogány. After her treatment was over, they all sat down to exchange a few words. László was to learn that Pogány, a divorcee with a son, Harry Kleinlein, living in Switzerland, was about to marry a Hungarian businessman, Lajos Láng, who had settled in Argentina. She was, as it happened, getting her teeth done in preparation for her departure.

"I have no idea what they charge in Argentina, you see—I don't want to start my new life spending hundreds of dollars at the dentist," she said. Looking at the two brothers, she continued, "Why don't you come? You want to wait until the Germans are holding knives to your throats?"

Bíró shrugged his shoulders:

"They won't give me a visa, and even if they did, I couldn't afford the journey."

"I was under the impression your family was well off."

"Indeed it was, a long time ago. But we've spent our every bean on a new invention, a new kind of pen."

"Can I have a look?"

Bíró showed her the prototype that happened to be on him and explained how it worked. Mrs Pogány seemed interested.

"I have a feeling something could come of this, you know. I have a habit of being right about such things. I would be happy to help you both. I'll make you a deal: we set up a joint venture out in Argentina, my husband and I finance the experiments, and in return we get the rights to produce and sell the pen in South America. Now there's a deal that's good for both of us."

Bíró nodded his approval, but could not imagine the offer was meant seriously; it was only out of politeness that he pocketed Mrs.

Pogány's business card, on which she had written her Argentinean address. He was more than a little surprised when, the next time they met, she produced a fully-blown contract.

"About as much use as a pile of Russian railway bonds," the inventor thought to himself, but figured that he did not have anything particular to lose, signed the contract, and forgot about the whole business.

What energies he still had, Bíró invested in finding a suitable financial backer, who might support the costs of perfecting the ballpoint pen in return for a percentage of the expected proceeds. His attempts were uniform in their failure. That Bíró was undergoing a bad phase of his life was reflected in the way he dressed: the trickle of funds flowing from Imre Gellért was barely enough for him to purchase a new shirt every now and then, and it had been years since he had last had clothes made. One problem brought about another. His wife was sent to a sanatorium with an apical lesion, while his daughter Mariann was being looked after by her grandparents. Bíró lost his enthusiasm for work, spending increasing amounts of his time at the Japan Café. He only ever had the means for a single mélange, but his old friend the waiter was not averse to adding the occasional "improving splash" of milk or of coffee, on the house, making his one drink almost as nutritious as a decent-sized meal. Leaning over the table with his head in his hands, he would stare into his cup for hours on end: he could not help feeling that he had ruined his life when he began dealing with the ballpoint pen, that whatever he might have done instead, he would have fared better.

On March 5, 1938, Bíró woke up in a bad mood. He turned on the radio to catch Prime Minister Kálmán Darányi's speech in Győr announcing that the country must devote all of its energies to the process of rearmament. When this was interrupted by the phone ringing, he had difficulty bringing himself to pick it up. It was Gellért.

"Listen, come to the Japan, right now, I think we may have found our guardian angel."

"That makes this an anniversary."

"What anniversary?"

"This is our hundredth offer. What's his name?"

"Goy, Andor Goy. That's right, you heard me. Goy by name, Goy by nature. His half-brother Dezső Gábor is artistic director at the theater where my wife performs. It was he who recommended us. He's a serious industrialist; a hundred people work at his typewriter repair workshop."

"You want to know the truth? I'm really not up to yet another failure."

"Come on, Laci, you're a good enough roulette player to know that you can't have the same color repeating itself forever. After a hundred black, you have to get a red. Pull yourself together and we'll meet in an hour."

CHAPTER 6

L et us introduce the second hero of our story. Andor Rezső Boldizsár Goy was born in 1896, three years before Bíró. An illegitimate child, his mother Ilona Goy was made pregnant by the "young gentleman" at the house in which she worked as a needlewoman. Goy makes no mention of his biological father's name in his memoir. His mother later married an alcoholic journalist of Swabian German descent called Emil Weiss. Weiss was the owner and also the whole workforce of the *Restaurateurs' Journal*; he would prise together the latest edition from various newspapers and the releases of the news agency using scissors and gum Arabic.

Goy's mother died after only a few years of married life; the young Goy was raised by Weiss's second wife under the most poverty-stricken of conditions. The boy made his contribution to the family's maintenance by doing shopping errands for the neighbors in return for small change. During the summer, his only source of entertainment was to plod along after the watering cart, washing his feet in the lines of water it left behind.

Goy showed signs of intellectual potential from an early age, and he was possessed with a persevering desire to achieve something with his life. He was not in a position to go to a normal grammar school, but he had a stroke of good luck. Thanks to the intervention of his step-father, he received a scholarship to attend Vilmos Braun's non-denominational private school. Goy paid for his accommodation by performing cleaning duties and various written tasks; in his free time he would cut up iron files and wrap them in paper.

After a year his scholarship ran out. The family's financial situation had not improved; moreover, Weiss's health continued to deteriorate. There was no possibility for Goy to continue his education. In 1910, five weeks before his fourteenth birthday, he became an apprentice typewriter technician. This profession probably attracted him because it required some intellectual skill and represented a cross between manual labor and technical fields. It may have been similar considerations that crossed the mind of another ambitious young man, János Kádár, who would become Hungary's leader for more than three decades during the communist era, when he opted for the very same profession some twenty years later.

Goy got up at five o'clock every morning in order to arrive on time at the workshop in the center of Budapest, on foot, from the village of Erzsébetfalva on the outskirts of the city. Before work started he would clean up and bring breakfast for the assistant technicians from the coffee-shop next door; he was not willing, however, to clean his boss' shoes, or to carry the repaired typewriters slung around his neck all across the city.

"I came here as an apprentice technician, not as a workman," he said. He would have let his sensitivity get him sacked, but there was no need for this, for his superiors appreciated his skill and diligence.

The boy was fast in acquiring the skills of his profession. By his second year as an apprentice, he was already commissioned to perform on-site repair and maintenance work. He earned enough to move to a separate flat with his step-mother, who had since been widowed. It was a humble affair, admittedly: a ground-floor flat with a single stone-floored room on the corner of Archduke Sándor Street, but it still represented a great improvement upon their earlier maid's room bed-sit.

The ambitious young apprentice could have done more to further himself, but on January 15, 1916, he was drafted. Many years later Goy's daughters drew up a table comparing the paths of their father's and Bíró's lives. They began by noting that during World War I, Bíró stayed in Budapest with a wound stripe while their father was sent to the Balkan front.

Goy was assigned to a technical unit, serving at a high-altitude cableway. He was sent home three years later, after being injured in an accident exactly when the communists took power and declared the Hungarian Soviet Republic.

Back in Hungary Goy had to start his life again from scratch. Lacking suitable civilian clothes, for a long time he continued to wear his uniform, robbed of its military emblems. He found employment at a commissariat office as a typewriter repair technician. After the short-lived communist regime collapsed, he applied for a license to trade independently. He divided off part of the one-room apartment as his workshop, separated by a curtain of torn bed linen. He installed a long table, which was a present from a gold-platter and frame-maker who also lived in the house. Goy affixed a wooden board to the wall where his tools were to hang; cleaning bowls were littered under his desk. It was in this most Spartan of environments that he began to work.

As a decommissioned soldier, he was initially well supplied with work from the Center for Former Servicemen, but as his reputation for careful and precise work spread, his client-base expanded, and in due course was unable to satisfy the demand on his own. He took on a young technician, István Kovalszky, as his partner; they would remain colleagues for decades, only to be parted by the nationalization of their factory twenty-eight years later.

In the spring of 1921 a well-dressed gentleman walked into his workshop holding a camera in which one particular component, the so-called "Maltese cross," had become so used as to be unserviceable. He requested the instrument makers to construct a new one just like it. Goy and his colleagues did not, as a rule, accept commissions to produce new components, but partly as a favor, and partly out of pro-fessional interest, they successfully undertook the assignment.

They accepted no remuneration for their work on the camera. The stranger, touched by their generosity, returned a couple of weeks later with an offer for them. It transpired that his father, Andor Juhász, was the chief justice of Hungary's supreme court. Thanks to him, Goy was offered a commission for the maintenance and repair of the typewriters used in Hungary's law courts. This significant and dependable stream of work allowed them to move from the curtained-off half-room to a real workshop. Fortune continued to favor Goy. For a long while after World War I the Hungarian market for typewriters remained in the doldrums, with no imports; the small ad columns of newspapers were full to the brim with potential buyers. It was in this tough climate that an agent by the name of Manó Komlós acquired some useful information. In the United States typewriter manufacturers accepted old machines in part-exchange when selling new models. Most of these trade-ins were in excellent condition, but it was impossible to sell them within the given industrial sector in America, and so they were exported at rock-bottom prices.

Komlós applied for a license to bring such machines into the country, using his relatives in the United States as intermediaries. The authorities, aware of the lack of supply in the market, supported his application, levying unusually low customs duties of sixty crowns per hundred kilos instead of the usual hundred and twenty crowns per unit.

Were a more sizable shipment to arrive, Komlós would send some of the typewriters to Goy, who would have only to clean and oil them and furnish them with new ink ribbons and keys with Hungarian accents before sending them out to customers.

Later, on the basis of a percentage agreement with Komlós, Goy was to establish direct contact with the Regal Typewriter Company in America. This was only the first in a long list of successful cooperative ventures: within a few years "Goy and Kovalszky" were the official Hungarian distributors of the most prestigious typewriter brands, including Underwood, Ideal, Erika, and the German Wanderer Werke Continental.

The typewriter repair workshop soon had more than a hundred employees, but as Goy was fortunate enough to have reliable technical directors in place, he could allow himself time to indulge his ambition to be an inventor. In the spring of 1938 he busied himself with the development of the design of a new, "automatic carriage return" take on the good old typewriter.

CHAPTER 7

In his memoir, Bíró cites Goy's name a total of four or five times; of their first meeting on March 5, 1938, he makes no mention whatsoever. Goy, on the other hand, gives a detailed account of what happened, both in his notes and in later interviews to newspapers.

He recalls that on that critical day it was he who arrived later at the Japan Café. Bíró and Gellért were waiting for Goy at a table overlooking Franz Liszt Square, where they were showing their ballpoint pen to other café regulars. As usual, however, the prototype was not working satisfactorily: their hands were covered in ink. They greeted Goy as he entered, introduced themselves, then wiped the pen clean with a napkin and handed it to him. The entrepreneur was not in the mood to negotiate in the midst of a throng of unsolicited onlookers, and invited the pair to come to his office at ten o'clock the following morning.

It dawned on Goy that this was not the first time he had come across such a pen. In his safe at home was the set of six Mungo ballpoint pens he had bought at the summer trade fair in Leipzig, Germany in 1932. He had originally intended them as presents, but by the time he was back in Budapest they no longer worked. He cleansed them thoroughly, and filled their cartridges with fresh ink, but by morning they were inoperative again, and he put them away.

In the safe Goy also found a leaflet on which he had scribbled at the time: "This is a great idea with fantastic potential." He took the box with him to the meeting, putting it down on the table in front of him.

"What is your profession?" he asked Bíró.

"I'm a journalist."

"An outsider, then. Where did you get the idea of working on the ballpoint pen?"

'The pioneers in the history of writing instruments have always been outsiders like me, not professional craftsmen. Waterman was a sales agent, who lost a profitable contract because of a pen that didn't work well; this is what encouraged him to invent the pen that bears his name. Parker was a high school teacher who wanted to make life easier for his students with his invention."

"Let's get down to business. Before we get into any details, I want to know whether you have patented your design for the pen yet."

Gellért shook his head.

"And why not?"

"You know as well as anyone the bill a patent application can run up. I have spent so much on this invention I simply didn't have enough left."

"An invention, did you say? Are you aware that the ballpoint pen was invented years ago?"

"You mean Wenzel-Klimes, the Czech? We know of him. The patent exists only on paper; he has no hope of putting it into practice. We can buy an option off him for a pengő or two any time we need it."

"That's not what I mean, gentlemen," said Goy, taking one of the Mungo pens out of the box. Bíró asked his permission to have a look at it. He couldn't make it work, either, but he was able to work out the principle behind it.

"This really is a ballpoint pen. Judging from the stains on it, it worked with ink. The acid in the ink has eaten away the ball and the socket, and the pen has become totally useless. The whole point of my invention is that is uses a special mixture of dye paste instead of ink. And that is adequate justification for a patent."

Goy furrowed his eyebrows.

"That's not for me to decide. We'll leave that to the patent agent I usually work with. He will do the research and submit the necessary registrations—on my tab, of course. In return I would ask for the option to experiment with your pen for two months. After that, I'll decide whether I am interested or not."

Goy became increasingly impressed with the ballpoint project, but his patent agent was less enthusiastic.

"There is nothing to stop us registering the patent, even if it has a certain 'parentage'; others have had the idea before, I mean. The patent office has no test for originality. They use what is called the 'Latin system,' where the granting of a patent does not, of itself, guarantee the invention's existence, worth, or uniqueness. The way the system tries to avoid abuses is by giving two months from the publication of the patent for anyone to appeal, if they can argue that it damages their interests. They can sue for possession, and apply for the patent to be made void."

"So what's the problem?" asked Goy.

"The problem lies with the character of the inventor himself. I must state my doubts concerning Mr. Bíró: he is a most ambitious fellow; indeed, it is possible that we have a genius on our hands here, but he does not have the necessary technical training and expertise. That can lead to really big problems. I'm afraid I cannot advise you to invest work, creative energy, and a lot of money in this affair."

Despite the agent's warning, Goy continued to see a great opportunity in the ballpoint pen. He commissioned his agent to register the patent.

I succeeded in getting my hands on the original document. Now yellowed and brittle, it states that on April 25, 1938, the journalist László Bíró registered his fountain pen at the Royal Hungarian Patent Office. His description was as follows: "A small ball embedded in a writing instrument such that it can gyrate, characterized by a cavity behind the ball suitable for holding a paste-like dye."

This was followed by a further Bíró patent on May 26 of the same year: "A replaceable ink cartridge for use with a writing instrument, and the writing instrument such a cartridge would be used in."

It continues: "The objective of the invention is the construction of an ink cartridge that makes it possible for anyone, in a simple way, to replace an empty cartridge with a new, full one, and further that these full cartridges be suitable for distribution to consumers." This would seem to signify the invention of replaceable ballpoint cartridges.

A few days later the two parties took their seats at the negotiating table again. Goy began by showing them the prototype, to which he had made one or two minor improvements.

"Have a look, gentlemen, give it a try."

Gellért picked it up and jotted in his notebook: "The ballpoint pen, the writing instrument of the twentieth century, sets off on its voyage of conquest."

"Congratulations!" he added.

"It's too early for that," Goy replied. "The dye is still a long way from being perfect, and there are problems with the cartridge, too. But we never got round to discussing your conditions for our possible cooperation. What are they exactly?"

Bíró instinctively turned to Gellért.

"I'll let you do the talking, Imre!"

"Shouldn't we involve a lawyer in this?" Gellért began.

"If you insist, we can," said Goy, "but, for my part, I'd be only too happy to do without. We industrialists have a saying: 'If the wind blows your hat into a lawyer's backyard, don't run in after it, for before you know it, it'll blow the money out of your pocket, too.' I think it would be more than satisfactory if Mr. Gellért were to speak on the inventor's behalf and my partner, Mr. Kovalszky, on mine."

"As you wish, Mr. Goy."

Gellért had another look through the notes in which they had listed their conditions.

"Our proposal was that our partner would, under an exclusive license, begin production and continue to produce the pen in the future. This would include foreign markets as well as Hungary. Let's begin with Germany: we are led to believe that you have very good contacts there. We could add Switzerland, Poland—assuming the Germans do not occupy it before then."

"I would ask you not to mix politics into our business affairs," Goy interrupted brusquely.

"Sorry, it is just hard for me to get this fear out of my mind."

"Let's move on, shall we? I would like to include France, too. I have two good friends living in Paris: Sándor Kovács and Van der Eynde, both engineers, who have a precision instrument factory under the name of Evelco. If we were to begin production, we would do well to get them involved. Do you speak any foreign languages, Mr. Bíró?"

The inventor smiled.

"Nowadays they tend to ask the question in a different way: 'Are you Aryan, or are you learning English?' I'm afraid I only know German, and only a little."

"It wouldn't be a bad thing if you tried to pick up some French. It may well be that in due course you will be in touch with the Evelco company."

In the end they made a list of some eleven European countries. Bíró's plan was to give Goy exclusive production and distribution rights in return for dividing profits as follows: 60% for Goy and 40% for them in Hungary, and on a fifty-fifty basis in the other countries.

"Would this same deal hold for other parts of the world?" asked Goy.

"It is possible that we would be willing to add a European country or two, but we wish to deal with the overseas markets, especially the United States, ourselves."

"Agreed. What are your financial needs, Mr. Bíró?"

"What were you going to suggest?"

"I am used to the seller naming his price, not the buyer."

"I was thinking of a sizable advance, to be deducted from my share of the profits at a later date."

"And just how sizable an advance did you have in mind?"

"Twenty-five thousand pengős. Five thousand in cash when we sign the contract, another five thousand when we present the formula for the dye in sixty days at the latest—I am sure my brother will be finished by then. And the remaining fifteen thousand by the end of the year."

There was an awkward silence. It was Kovalszky, who had only observed the conversation thus far, who broke it.

"Mr. Bíró, that's rather a high price, don't you think? Twenty-five thousand pengős is more than five thousand dollars; that's enough to buy a nice house. Yet, you hardly have anything to offer us in return. The future of this ballpoint pen is completely uncertain."

It was Gellért, the experienced deal-maker, who responded in Bíró's defense:

"How right you are! We are indeed talking about a lot of money here. But this is a stake you're putting on a horse that has a fighting chance of winning the derby—especially in the hands of jockeys like your good selves."

Everyone waited for the final word from Goy; it was he, after all, who held the purse-strings. He did not give one, not yet.

"I would like to ask for two more days to think over your offer. I will give you a final answer in the last days of May, let's say the thirtieth."

Bíró could not help being reminded of the contract he had signed with General Motors. He wanted to be certain that Goy would not bury his invention forever.

"I have one more condition, Mr. Goy. I would like the contract to state that production will begin no later than December 31, 1938."

"That's a pretty tight deadline, but I think I can live with it. Good day to you, gentlemen."

CHAPTER 8

The contract was signed on May 30, 1938. Once they had stepped outside onto the street, Bíró, the first five-thousand pengős in his pocket, put his arm around Gellért.

"You are the one I have to thank for everything, Imre. You didn't wait till midday before you started believing in the sun," he said, trying to stuff a wad of notes into his friend's pocket, "Here, this is the first installment."

Gellért pushed his hand away.

"Don't be ridiculous. You're the one who needs that money. But I do think we deserve to take a look in Mr. Staub's tailor's shop on Semmelweis Street and have ourselves a decent suit fitted."

"I've never dared to go there, I have to admit. I'm told you can spend as much as a thousand pengő."

"They're not cheap there, no question about that, but as they say in German, *qualität kostet geld*, old man. I haven't been there for more than two years myself. Sometimes I run into Mr. Staub on the street, and he always asks me, 'Business not going so well? Bit down on our luck, are we, Sir? Don't worry, Mr. Gellért, you can pay at your pleasure.' But I don't like to dress up on credit, you know."

As they stepped into the shop, the tailor left his client in the fitting booth and hurried to greet them.

"Mr. Gellért, if only you knew how delighted I am to see you. This does mean, I hope, that your business is blossoming once more?"

"My travel agency continues to be as popular as a matzah stall outside the Vatican, I'm afraid, but my good friend has hit gold. Have a good look at Mr. Bíró here, he's the one who'll be coming here twice a year from now on."

"Can I be so bold as to ask Mr. Bíró's line of trade?"

Gellért took the ballpoint pen out of his pocket.

"This is his invention. It is going to make him a millionaire. Why don't you try it out, Mr. Staub?"

Expressions of impatience could be heard from the fitting booth. As a means of excusing himself, the tailor asked if he could show the pen to his client who was waiting inside. Moments later the curtain was drawn back to reveal a black-haired man with a mustache and a gleaming set of teeth, wearing a sleeveless jacket with fitting marks, and with the ballpoint pen in his hand.

"I would be very interested in this pen. I'm sorry, we have not been introduced. I'm Vili Víg, from Belgrade. I work in foreign trade, and I have an agency on Terazija Boulevard." He looked at his watch. "My wife is waiting for me at the Keleti train station, so I will get to the point: do you have an agent for this product on the Balkan Peninsula?"

Gellért reacted instinctively, attempting to improve their bargaining position with a little bluff:

"We have had a number of offers, but have not signed any contract as yet."

"Then wait before you do sign one," said Víg, getting dressed. "Let's discuss this matter in more comfortable surroundings. If I may make a suggestion, I am going on holiday to the seaside in July. I would like you to be my guests, let's say from the twenty-eighth. Please make note of the date and of the name of the hotel: Hotel Jugoslavinka-Rochlitz, or, to use its new name, Roganska Slatina. Could you give me a business card, so I know where to send the train tickets? I would be glad if you could bring your wives with you; my wife is from Budapest too, she would enjoy their company."

"That won't be possible, I'm afraid," said Bíró. "My wife is in a sanatorium, while my friend's is an actress, on stage every night."

"I am very sorry to hear that, but I look forward to seeing the two of you in any case," said Víg, waving goodbye and getting into his taxi. Bíró and Gellért stood there, a little stunned by this sudden barrage of words, watching him leave.

"That's what they call the American tempo, I suppose; he floored us in five minutes. Mr. Staub, what do you know about this gentleman? Not exactly short of a few bucks, I suppose?"

"He's never made any attempt to show that he is wealthy, so he probably is."

Víg sent them the train tickets as he had promised. Bíró and Gellért traveled down to Roganska Slatina, where the suite was waiting for them at the hotel. Gellért tried to ask for something smaller.

"Better to be careful," he advised Bíró. "Víg is nowhere to be seen, maybe he won't even come, and we'll be left here washing up to cover the cost of the room. At least with a cheaper room we won't be stuck here for quite so long." They thought that a little bit of everyday propaganda couldn't hurt, and so they filled in the registration form with one of the display models of the pen. Beside them at the reception desk stood a bespectacled man of about sixty, in the company of his young secretary. He observed the unusual item with interest. He spoke in Spanish, and it was with the porter's help that he expressed his wish to make the Hungarians' acquaintance.

"General Augustin P. Justo would be glad to receive you in his apartment between 4 and 4:30 this afternoon."

Gellért accepted the invitation with a respectful nod of his head, but did not take it particularly seriously.

"Do you know how many generals there are in Argentina? More than there are privates! I read it somewhere. Anyone who knows how to tango is made a general."

Bíró shook his head.

"Doesn't cost us anything to meet him."

Víg finally made his appearance. Inviting them to join him at his table, he wasted no time, asking to look at the pen during their lunch. He studied it carefully.

"How do your plans for mass production of the pen stand?"

"We have plans, but everything hinges on whether we can find the start-up capital. We already have a contract with a Hungarian company that will produce and distribute the pen, but we would also like to establish contacts elsewhere."

"So you are still at an early stage in all this. No problem, I rather like it when the starting-pistol fires and more depends on the rider than on the horse. I may be the very man you are looking for. My retail network covers the whole of the Balkans, and my bank account is buzzing with the flow of transactions." Víg raised his hand. "I won't be at all insulted if you run a check on that."

"What do you propose?"

"First I would like to sign an option contract to act as lead agent for the ballpoint pen in the Balkan countries. This should not mention any price or time limit. This product is still in flux: it might one day be worth its weight in gold, and it might equally hit the scrap heap with a bang. I myself think it has a great future, the idea is such a good one."

Gellért interrupted him.

"As far as I'm concerned a compliment is only a compliment if it's expressed in figures. I'll make myself clear: what are you offering us for this option contract?"

"Forgive me, but that question is not fit for a professional. You are asking for money for first refusal on a product that to all intents and purposes does not exist. I should be glad you're not trying to hustle me into buying a perpetual motion machine at a discount!" Víg gave a sigh, as if to signal that his enthusiasm had been quashed irreparably. "If you'll excuse me, I have been traveling all day, and I would like to go up to my room for a nap. We can continue our discussion in the evening."

Gellért watched him leave with a grimace.

"Not another one. He'd take one look into the water and the fish would bite the dust in seconds."

As far as Bíró was concerned, the best thing to do would be to wind down with a stroll by the sea, but Gellért rapped the glass of his watch with his finger.

"It's past four—let's not keep our little general waiting."

They went up to Justo's suite, and Gellért looked around in some astonishment.

"You wonder how old Justo will pay the bill. I have a feeling he might end up doing the dishes with us."

General Justo emerged from the depths of his room with a friendly smile on his face.

"I hear you are Hungarians. I am very fond of that little country. I know you have given the world a good number of priceless inventions. I really hope that this ballpoint pen will be one of them. Argentina is the place for such talented men as yourselves. I am sure a glittering future would await you there."

"It's not as simple as that, I'm afraid, General. They have stopped giving out entry visas to citizens of Central European countries. The ship is full, so they say. It's not as if we haven't tried."

"I may be able to help you there. But I think there is a much more promising option."

"What would that be?"

"I'll be honest with you: I am no expert in day-to-day business matters, neither is it my job to be. I have the necessary advisers if I need them. But I have the feeling that our industrial development program could use this ballpoint pen idea: limited raw materials, high added value. We could set up a lobby to press for the realization of this invention in Argentina. Only if you both agree with the idea, of course."

"We would be delighted."

"I am leaving tomorrow morning, so we need to arrange our next meeting."

The general asked his secretary to give him his calendar, and looked through it at length.

"This is my last trip to Europe this year, the next one will be in the spring. What about meeting on the tenth of March, 1939, at the Argentine embassy in Paris? Say ten a.m.?"

Gellért whispered to his friend:

"That the biggest bluff I've ever heard. He picks out a date just like that, eight months in advance!"

Bíró tried to calm him down.

"Try to behave!" He turned to the general. "Unless anything unexpected happens, we'll be there. Of course, we will have to discuss the matter with our partners, but I really hope that nothing will prevent our meeting. As far as expenses are concerned . . ."

"Don't worry, I will take care of all that."

General Justo took out a business card and scribbled a couple of lines on the back.

"There you are. I have noted down the time and place of the meeting, just in case you forget. You should show that card at the entrance to the embassy. The visas will be waiting for you at the Argentine embassy in Budapest. You will also be told which hotel I have booked for you in Paris. It has been a pleasure meeting you, gentlemen."

They made their way down the stairs at a slow pace. Bíró was in a contemplative mood.

"Don't you find it strange that Argentina has come up again in such a short space of time? First Mária Pogány, now this general chap."

"Surely you didn't take all this at face value?"

"Why not? There's no sin in holding out hope."

"I wish you were right, Laci. I'm afraid that at our age it's a sin like any other. You can have hope for breakfast, but not for dinner."

"Whatever the case, we will go home in the morning and work our fingers to the bone a bit more. Come on, let's go for a walk."

On the promenade, they ran into Víg. He invited them to join him for a glass of wine in the pavilion by the beach.

"No hard feelings, I hope? You're not angry that I didn't offer to buy the option?"

"Why would we be? There are no emotions in business."

"How did you spend the afternoon? I imagine you had a bit of rest."

"We didn't have time, actually. We met a gentleman from Argentina to discuss the production of the pen there."

Víg looked up.

"From Argentina? You never told me about that. Could I ask for the gentleman's name?"

"But of course. This is his business card."

"He gave it to you himself?"

"Yes. We agreed to a meeting in Paris."

Víg took a glance at the card, and for a moment or two was at a loss for words.

"Do you know who you were talking to?"

"He said he was some kind of general."

"Augustin P. Justo was President of the Republic of Argentina for twenty years. At the moment he is on a semi-official visit to Yugoslavia to negotiate a multimillion dollar agreement between the two countries. If he judged you worthy of his confidence, you have struck gold. Let's celebrate with a bottle of French champagne!"

As the waiter filled their glasses, Víg gave Bíró a couple of brotherly slaps on the shoulder.

"During my siesta I had a chance to rethink this matter. In the first instance I would need forty thousand pens and a fixed contract, with payment in cash. As for the option, I would be willing to risk a bit of money. Don't expect riches, but enough for you both a few more suits from our favorite tailor, Mr. Staub. Another bottle, gentlemen?"

It was after midnight before they finally left their table. No sooner had the two Hungarians sat down on a moonlit bench to clear their heads, than Bíró began to indulge in a moment of philosophizing.

"Imre, have you thought about how many coincidences we needed for all of this to happen? It was because we got the money from Goy that we went to the tailor's at the exact time that Víg was trying on his suit. He invites us down here, then we run into a former President of Argentina."

"We need to concentrate on business, not on the quirks of fate. Víg said that he would order forty thousand pens. How many can we sell to the Argentines?"

"What's the population of Argentina?"

"Don't know, fifteen or twenty million, I would think."

"Then a hundred thousand, at least, and that's just the beginning. And if we establish a foothold in Argentina, we will soon have the whole American continent in our hands."

"And that's not to mention Australia, Asia, Africa. More than a million pens a year."

"That's enough for now. Let's not get carried away!"

As Bíró was to write a few decades later, "In retrospect I have to laugh at what I thought was my 'nerve.' Today a single company in Argentina produces two million pens a month." This figure has only increased since.

CHAPTER 9

A t the beginning of summer in 1938, Goy planned a business tour of Germany. He wished to continue negotiations about the distribution of his invention, the automatic carriage return typewriter, and to use the opportunity to show the ballpoint pen to his business partners.

Despite working on the experimental models of the pen around the clock, Goy still did not find their technical quality satisfactory. His attaché case and his pockets were brimming with pens, and he spent his every spare minute trying them out, seeing whether they really would being writing at exactly the moment they touched the paper. Only too often he would find that they were slow in starting to write; later the lines would often break up like streams of Morse code. Goy began not only to fear for his professional reputation, but that he would suffer disaster at the German presentations that were fast approaching. He would work until late at night, and no matter how much he would try to give up work for the day, the adrenalin continued to flow. He would often wake up in the middle of the night, pull his overalls over his pajamas and return to his workshop, experimenting with potential solutions until early in the morning.

On one occasion, attempting to explain the "gaps" in the flow of the writing, he inspected the way the ink was supplied to the ballpoint. At the bottom of the ball socket was a hole half a millimeter wide through which the paste flowed from the cartridge toward the pen's tip. It occurred to Goy that this hole might not be wide enough. With many years of experience behind him, Goy set out on the exper-

GYÖRGY MOLDOVA

iment with confidence. He was to write:

> I took out a little three-cornered pinhead file, grinding the tip of the file until
> it was smooth on all three sides. I lifted the ballpoint from the tip of one of
> the pens, and softly pushed the file through the highest ridge of the 0.5 mil-
> limeter bore hole, creating three tiny recesses, hardly visible to the naked eye,
> that would hold ink.

> Afterward I replaced the ball in its socket, resealing the part of the pen that
> keeps it in place, and screwed the writing tip into a cartridge full of ink. I
> turned on the pump that pushed the dye paste, and tried out the pen, which
> immediately began to write in a continuous line.

Although this idea played an important, possibly definitive, role
in the perfection of the ballpoint pen, Goy was filled with professional pride in the skill he had displayed, not with a sense of urgency
to register a patent. At the bottom of his heart, he did not really think
that the pen would become a popular product around the world; he
had greater hopes for his other invention, the automatic carriage
return typewriter. Later, when others profited from the discovery he
had made, Goy would come to regret his inaction.

With a dozen pens that had been perfected in this way in his
bag, Goy set off on his trip around Germany. His first stop was Sieg-
mar-Schönau, where he had meetings about his typewriter invention.
These proved successful; the German firm signed an option contract
for the machine, and purchased it five months later.

Following this, Goy concentrated his attention on the ballpoint
pen. In Nuremberg, he stopped by the famous Faber-Castell factory,
saying he wanted to present a new pen based on a revolutionary principle. The technical director, who introduced himself as Herr Sattmann, was willing to receive him.

After they had dispensed with the necessary formalities, Goy
took the pen out of his attaché case.

"If it doesn't work now, I'm sunk," he mumbled to himself.

He drew a couple of lines to test it, was relieved to find that it
worked, then wrote these words on a piece of paper: "*Hiermit wird
Ihnen das erste Kugelschreiber muster vorgeführt.*" He placed the paper

and the pen on the table in front of Herr Sattmann.

"There you are!"

The German looked at the pen reluctantly, gave it a quick try, but replaced it almost at once.

"If you are under the impression that this is a new invention, you are much mistaken. In our archive we have a whole list of patent registrations on this subject, from Germany and abroad."

"But none of them have ever made it to the shops!"

"Do you think that is some sort of accident? This crude instrument is no competition for a classical steel- or gold-nibbed fountain pen."

Goy was reminded of his own debate with Gellért and Bíró, in which he had been the one playing devil's advocate, and tried to reconstruct what the other side had said:

"It appears that I haven't made myself properly understood. This particular ballpoint pen has a score of advantages over earlier versions. It doesn't work with ink but with a paint dye, and it doesn't spill, but writes continuously. Not to mention the fact that a traditional fountain pen costs a fortune, while this pen can be sold for the price of a pencil."

Even these arguments failed to convince Sattmann. He did not want to get caught up in a lengthy debate, and asked his guest to leave a few prototypes behind, promising that he would inform him of his decision within a few days.

On the next stop of his journey, the Pelikan factory in Hannover, Goy did not even get this far. The director of the research laboratory had scarcely picked up the pen when he returned it to him without even trying it.

"My good Sir, this has been a waste of your time, I'm afraid. There's nothing special about this."

Goy traveled on from Hannover to Hamburg to try his luck at the Mont Blanc factory. He did not meet with any greater success, only managing to leave some samples for them to test. A few days later he made a call to Budapest. Kovalszky reported that both Pelikan and Mont Blanc had returned the samples, with nothing more than a cover letter saying that they were not interested. But there was a glimmer of hope: DFW, Deutsche-Amerikanische Füllhalterwerke, had expressed an interest in meeting them. Goy decided to give it a try.

Munich, the cradle of the Nazi movement, was where DFW was located. Herr Zorn, manager of the factory, was waiting for Goy at the train station with a swastika on his arm. He greeted his visitor with a Nazi salute, then, after giving him a tour of the city, took him to a political meeting being held in a bierhaus. Goy, who always had a good appetite, enjoyed the "standard menu" of sausage with boiled cabbage and potatoes served with a half-liter jar of beer. He felt threatened by the speeches and the aggressive emotions of the participants, however. He was only too relieved when they finally left the hall. Goy read the newspapers and kept abreast of world events, but this was the first time that he had seen Germany so obviously gearing up for war with his own eyes. He was afraid that Hungary would also be drawn into the trouble they were concocting.

The following morning Zorn took him to the DFW factory, where the board of directors welcomed him. Goy showed them the ballpoint pen in his usual manner and was delighted to see that, for once, his words were not falling on deaf ears. On the contrary, the members of the board were giving him their undivided attention. They passed the pen around between themselves, all giving it a try. It was the leader of the patent division who gave his opinion first:

"The German people are never resistant to innovation. In this, as in so many other things, we are superior to the English or the French, but its decisions always attest to a healthy skepticism. Our people will not change their mind overnight, nor will they allow themselves to fall into the same trap twice. As you are well aware, the ballpoint pen once appeared on the German market under the Mungo brand. It was such a flop that newspapers actually published editorials telling people not to buy them. In such circumstances, we have to think twice before embarrassing ourselves with another crack at it."

The chairman of the board stopped the engineer in his tracks.

"Your eloquence is commendable, Mr. Baumann. You made a mistake in not opting for a career as an orator. On this occasion, though, you are wrong, because you fail to consider the most important factor, the present condition in which our company finds itself." He turned to Goy. "I think we can speak openly in front of you; we are sure you will not use the information against us. I am sure you know that we produce the famous Tintenkuli fountain pen for the Riepe Company in Hamburg—the pen with the red ring around it. Our contract with them expires at the end of the year. Riepe does not want to extend it—it has built its own production facilities in the meantime. This leaves us without a lead brand. Should this ballpoint pen be a success, and take the place of the Tintenkuli on our production line, then all I can say to you, Mr. Goy, is that you have been sent by God."

In the minutes afterward the atmosphere in the office began to resemble that of an arcade hall. The members of the board were all busy playing with the pen. Zorn started drawing lines and circles on a piece of paper held above his head.

"Why, with this you could even write in the stratosphere. Can I suggest that its name be Stratos?"

Zorn and his colleagues would have been willing to take out an option contract on the invention at once, but Goy, who wanted to gather information about their company first, asked for time to consider their offer. He urged his new business partners to test the pen as thoroughly as they could, carrying out whatever experiments they felt necessary, and to find out how a patent could be registered for it in Germany. They arrived at a verbal agreement that, should no obstacles be found, production would begin by December 20 at the latest.

Back home, Goy reflected on the results of his trip. His conclusion was a sobering one: three really prestigious companies had rejected his advances out of hand, which was a bad omen, even if it was compensated for by DFW's enthusiasm. But he decided that he had to take the project through to its conclusion.

Doing some research into the Munich company, Goy convinced himself that it was both technically and financially sound. Their cooperation began with Zorn and his colleagues setting out to produce a trial line of what was now definitively called the Stratos pen. The company's chemical engineer, Dr. Zychlinsky, was given the job of perfecting the dye paste.

The whole affair continued to face adverse political developments in the world outside. The German war machine was already making preparations to attack Poland. With the slogan "Guns, not butter," production of civilian goods was overshadowed by that of military ones. More and more often, DFW would report impediments to the introduction of the Stratos pen, the most serious of these being Dr. Zychlinsky's conscription into the army. Goy's friends in Paris, at the Evelco Company, were also pursuing research into the dye paste, but without any significant results.

In the light of all this, it became increasingly doubtful whether production could begin in 1938. Yet, a delay past the end of the year would contravene the contract signed with Bíró. Goy does not remember meeting the inventor on a regular basis; when Bíró came into his office, it was to inquire about the progress of the research, or to ask for some samples. If they ever did fall into conversation, Bíró would try to provide explanations as to why his brother had not yet succeeded in producing the promised dye paste formula, even though they had already received payment for it.

For the time being Goy made no attempt to amend his contract with Bíró; he only sat down to discuss matters with his partner Koval-szky. The latter had never understood why a company distributing and repairing typewriters should be involved in experiments on a ballpoint pen, but he had resisted the temptation to voice his objections. With the conditions now so much riskier, Goy thought it only fair of him to assume the whole burden of the ballpoint project himself and repay the costs hitherto incurred to the joint company from his own private finances.

In early December, Goy once again traveled to Munich and, despite the prevailing difficulties, he succeeded in signing an agreement with DFW.

There was no way that the original deadline could be kept. DFW agreed to begin production by July 1940 and to sell the first hundred thousand units by July 1941. Little was either side to know that by then, the German-Soviet war would be well underway.

CHAPTER 10

It was Gellért who passed the message to Bíró that Goy wished to meet him on the morning of December 23, 1938. Bíró had by then long determined that he would leave Hungary before the end of the year. The final shove that strengthened his resolve was the information he received about an impending item of legislation, to be put into effect from January 1, 1939, that would make it illegal for Hungarian intellectual property to leave the borders of the country. He was worried that this would, in due course, affect his invention. He had spent much of the last months going from one editorial office to the next, trying to cobble together a few hundred pengős from odd jobs, but not with much success.

Everyone was busy preparing for Christmas at home as Bíró set off for his meeting with Goy. The shops closed, and he saw little in the way of traffic as he rushed through the delicate mist. The same pleasant seasonal atmosphere prevailed in Goy's office on Nádor Street; the lights on the Christmas tree were already lit, and in the entrance hall his employees were enjoying a toast. Goy broke off from the crowd to greet Bíró with a firm handshake.

"Good to see you. Come and join us."

Goy would not normally think of letting anyone have a drink in the workshop, but to get into the Christmas spirit he fetched a bottle of cognac and some glasses. Looking a trifle bored, Bíró sat back in the leather armchair and glanced over at the papers on Goy's desk. He had his journalistic background to thank for the ability to read documents upside-down. He could not help noticing a letter with the letterhead of the German company Deutsche Füllhalterwerke, and he slowly deciphered its contents:

SWORN AFFIDAVIT.

I hereby accept, on oath, my responsibility to help the creation of a Jew-free Europe by refusing to sell the products of Deutsche Füllhalterwerke, whether directly or indirectly, to any Jewish company. I also accept responsibility to ask for declarations similar to this one from every one of my wholesale purchasers. Should there be evidence to show that I have not kept to the aforementioned obligations, DFW has the right to regard my orders as void.

Bíró imagined that the text of the affidavit had probably been worded by the German company and sent round to their business partners as a circular. As far as he could see, it had not yet been signed.

In the meantime, Goy had finished serving the brandy and lifted his glass to drink a toast.

"Merry Christmas!"

"Thank you, to you, too. And a very Happy New Year."

"That can wait, we'll see each other before then."

"Probably not. In a few days I am leaving the country. I have already been granted a passport."

Goy looked up.

"Are you going away on business?"

"No, for personal reasons. I don't feel like waiting for the Germans to invade the country and make me wear a yellow star."

"I don't think you need to be quite so pessimistic. The Germans' bark is worse than their bite."

Bíró shrugged his shoulders.

"You may be right, but if I get it wrong, the stakes are deadly."

Goy was suddenly struck by the memory of the meeting he had witnessed in Munich.

"I can understand if you don't want to take the risk. Where will you go?"

"France. I will try to earn enough money to pay for my family to come and join me."

"Have you heard from that Argentine general yet?"

"I haven't, but that doesn't mean anything; our meeting is on the tenth of March, in Paris. Until then, I'll try to find work somewhere."

"I might be able to help. I'll write to my friend who owns Evelco and ask him to hire you. They may even provide a roof over your head."

"Thank you."

"When do you intend to return?"

"Probably never."

"I understand. I honestly hope that things don't come to that, that the situation here gets better, and you feel comfortable coming back. But before then we must settle a few business matters."

Goy pulled out a folder full of documents.

"There's something I have to tell you. I was in Germany, where I had meetings with DFW. I really wish I could, but there's no way I can start production of the pen before the end of this year. If you don't object, I would like to extend the deadline to June 1940."

"No problem. Are there any other amendments you would like to make?"

Goy thumbed through the pile of papers.

"There's no denying that the costs involved have been far higher than what we expected. I think it is only fair if we make a slight adjustment to the original division of the proceeds in my favor." Goy read through the figures in a monotone, and Bíró hardly interrupted him, except to insist that the status quo be kept for the French market, in the hope that there he would be able to access his money directly. When Goy had finished, Bíró nodded his head.

"You make the decisions as you see fit. The only thing I ask is that I be involved in these matters as little as possible."

"I suggest we avoid getting lawyers involved. With them, everything gets so complicated; time is not on our side, after all. You are the one who knows how to write. I'll just draw up an outline of the paragraphs involved, and you can do the actual wording. I suggest that between Christmas and New Year an errand boy should take the outline to you. You type it up and send it back to me."

"That is also fine, but I have one condition—that the errand boy shouldn't arrive empty-handed. He should come with more than just the sketched contract, I mean."

"What else should he bring?"

"I would like to ask for another five thousand pengős, to be deducted from my share of the profits later. I don't want to leave my family behind without a penny."

Goy frowned.

"It is not a businessman's job to take care of his partners' families. This is the first I've heard of this."

"It's also the first I've heard of the things you are asking for. Believe me, if I had twenty-five thousand pengő, I would excuse myself from the conditions of the contract and pull out of it altogether."

The industrialist pondered before nodding in agreement.

"All right, but I can't risk sending five thousand pengő with an errand boy. You'll have to come in and get it yourself."

"You're right. I won't trust him with such an important contract, either; I'll bring it in with me."

Goy laughed.

"Looks like our trust is mutual. I will tell accounts immediately, and you can pick up the money there. Another cognac?"

"You know what? I think I will."

"I have a feeling that if we hadn't met in the way we did, we might have become good friends."

"Let's not get into all that. Sentimentality is to the soul what hyperacidity is to the stomach."

At the time they drank their toast, Bíró was thirty-nine years of age, Goy forty-two. It did not cross their minds that this might be the last time they would ever meet.

As the letter that Bíró subsequently composed formed the basis of a multimillion dollar license trial, it seems right to quote it here in full:

Budapest, December 24, 1938

To the attention of the honorable András [sic!] Goy

Budapest

On this day we have mutually agreed to void our previous agreements concerning the so-called "Bíró" pen, choosing to enter the following agreement instead:

We establish a joint partnership with respect to the use of the pen in Hungary, as well as Germany, Yugoslavia, Romania, Greece, Switzerland, Poland, Estonia, Bulgaria, and Lithuania, according to the following conditions: in Hungary you are entitled to two-thirds of the net profits attained, and I to one third. The profits arising from sales in the aforementioned territories are to be divided evenly. As far as France is concerned, the conditions of our earlier agreement continue to hold.

In all countries outside those mentioned the distribution rights are mine alone.

As far as the sale of the product in Hungary is concerned, our plan is for you to set up a production line to satisfy the demands of the Hungarian market. In this instance, however, the material costs associated with the construction of the factory are not to be accounted for as general overhead costs. You are to provide the necessary funds in the form of an advance, which will be accounted for later in the form of amortization. You are to commission Füllfeder Fabrik A.G. to satisfy the production demands of other countries, from whom we are to receive a royalty in return for each item produced or sold, or with whom we are to settle our mutual interests in a different way.

In order to protect and advance our business interests, we believe it necessary to register the patents that you are acquainted with and that reside with our common agent in all of the aforementioned countries. Should it by some chance not prove to be possible to have the cost of this defrayed by our expected German or other partners, you have accepted to pay them yourself, in the form of an advance that will later be settled from our expected joint profits. In view of the fact that I am in receipt of a profits advance from you that is yet to be settled, we hereby agree that you will at all times withhold one third of the profit due to me as a partner and use it to redeem this debt of mine to you.

My fullest regards,

László Bíró

On behalf of Mr. Imre Gellért, I hereby state that he has also agreed to regard as void the former agreements that affected him, and so you bear no responsibility to him.

CHAPTER 11

Bíró insisted on keeping to his original plan of leaving Hungary before the end of 1938, but as he wanted to stay with his family as long as possible, he bought a train ticket to Paris on New Year's Eve. In the morning, the members of the family gathered to wish each other a Happy New Year, then Bíró retired to his room to pack his things. Gellért sat next to him, chainsmoking. From his desk drawer, the inventor pulled out the blue folder in which he collected his ideas, the notes from his experiments, and important business cards.

"You're taking that as well?" asked Gellért.

"Of course I am. This is my only possession in the whole world. You never know, maybe I'll meet someone who's interested in it."

"The border guards will catch you with it."

"You're right. But until now, every time I haven't taken a risk, I've regretted it. I'll stuff it between my clothes." He continued with his packing, his movements quite relaxed. "Maybe on New Year's Eve they won't be so strict at customs."

"You're good at hiding your fear."

"You know why? Because I'm not afraid. I am wearing my 'shirt painted in dragon blood', even now."

"What's that?"

"When I was about six or seven, my father gave me a shirt, a white shirt, with a thick red stripe across the chest. He told me that the stripe had been painted with dragon blood and that if I put it on,

I would be invincible. I really liked the sound of that, and I wanted to try out its magic powers at once. I went out on the street and hit a cyclist on purpose. He fell over and almost broke his neck, while I didn't suffer so much as a scratch. Then I stood up in the window of our apartment, gripped the frame between the open windows, and started jumping in and out. When I didn't fall out, I knew this shirt really made me invincible."

"You still have it?"

Bíró pointed to his chest.

"No, but I still have it in here."

Bíró closed his suitcase and took a seat. He drummed his fingers on the table nervously.

"If only I could get this departure out of the way. Once I'm on the train, I'm half way there."

"Is there anyone waiting for you in Paris?"

"I sent a telegram to Faragó asking him to book me a room for one night. The second of January is a workday, so I'll go and visit Evelco. Goy has friends there, he wrote me a reference." Suddenly he looked up at Gellért. "Are you sure you don't want to come?"

"Me? Come on, you know what a troublesome character I am. If I could leave myself at home, I would come, but I can't exactly come without myself, can I? I might as well stay here."

They took a taxi to the Keleti train station at eight o'clock. The streets were already crowded with the celebratory crowd. Bíró looked out of the car. He would have liked to pick out something he could forever remember the evening but didn't see anything of note. His six-year-old daughter Mariann sat in his lap and waved to the passers-by with the hand of her doll.

"Bye-bye! Daddy's saying bye-bye to Budapest!"

Bíró said farewell to his friends and relatives in turn; only his wife Elza boarded the train with him. The compartment was empty. Elza arranged her husband's belongings, then sat down on the seat opposite him.

"Another ten minutes to go. It would be nice if someone else joined you and you didn't have to be so alone."

"Who else would travel at a time like this? Take good care of my mother, she's really getting old now."

"I know, she keeps saying she 'can't help it.' Don't worry, my love, everything will turn out just fine."

There was silence. It was Elza who broke it.

"I miss you already. Yesterday I was thinking about it. Apart from when I was in the sanatorium, we've never lived apart for longer than a week, and even then we saw each other every day."

"You'll come and join me, maybe in the spring."

"Don't let your pride get to you! If it doesn't work out, just come home, we'll try something else."

"It has to work out."

When they announced that the train would leave in five minutes, Elza got up from her seat.

"I put a bottle of champagne in your suitcase. Break it open at midnight, and I'll do the same at home. It will be as if we were together."

"Go now, before it's too late. Take care of yourself."

"You take care of yourself."

Elza had hardly reached the platform when the guard blew the whistle. The signal was raised and the train pulled off. Soon the figure of Elza as she waved was hardly visible, then, as the train turned the corner, it disappeared completely. Bíró took the deepest breath of his life, terrified that his heart would stop beating.

He sat down, unbuttoned his shirt, and was relieved to find that his heartbeat had become stable again. The knowledge that he was now on his own gave him a renewed burst of energy. He took out his French textbook.

"*Pourquoi m'avouez vous, Monsieur? C'est mon petit travail, n'est-ce pas?*" he mumbled to himself.

The train stopped often as it crossed western Hungary to be cleared of the thickly falling snow, and it took eleven hours to make the 110-mile journey to the border at Hegyeshalom. Even if the border guards found the patent documents, Bíró thought, they could not punish him, as the ban on the export of Hungarian intellectual property would come into effect only at midnight. He was nevertheless much relieved when the customs officials, busy preparing for their midnight celebrations, let him through without any particular inspection.

What had been the Austrian border was guarded by soldiers dressed in the uniform of the Third Reich. They took a long look at Bíró's passport. He had converted many years earlier, and his religion was described as "Augustan Lutheran"; he considered this slightly less traitorous than becoming a Catholic. His conversion did not exempt him from the stipulations of the Jewish laws, but he felt it had helped his entry into the Reich. The guard endorsed his passport with a stamp; above the text was the figure of an eagle with its wings spread.

Once the train reached France, Bíró managed to sleep for a couple of hours. When he awoke, it had become distinctly milder, and the falling rain drummed its rhythm on the train roof. The ticket inspector opened the door of the compartment.

"C'est Paris!"

The glass hall of the Gare de l'Est was full of the haze from the exhaust fumes drifting in from the street. The lights were struggling to illuminate the gloom. A few lamps were all that bore tribute to the celebrations.

The inventor put down his suitcase and leaned against a column to wait; Faragó arrived a few minutes later. Bíró gave him a hug and a kiss on each cheek, he was so happy to see a familiar face and to be able to speak Hungarian again.

Faragó invited Bíró to the regular haunt of Hungarian émigrés in Paris, the Café des Deux Pistolets. While they sipped their coffee, he asked about the situation back in Hungary, and told him about the eminent Hungarians currently residing in the French capital, including Mihály Károlyi, the former President of the Republic, and Lajos Tihanyi, the painter. Bíró was delighted to hear that his friend László Lakatos occasionally made the trip to Paris from his residence in Nice.

Afterward, Faragó took his guest to where he would be staying, a little hotel on the Avenue MacMahon, and bid him farewell.

The following morning Bíró presented himself at the Evelco factory and showed them his letter of recommendation from Goy. It was with barely concealed reservation that he was welcomed by the two owners, Sándor Kovács and a Hungarian-born Belgian, Lucien van den Eynde. As Goy notes in his memoir, "they were not exactly anti-Semites, but they were not the greatest Jew-lovers, either." Bíró sensed their hostility, but he needed the job, and tried to put the discomfort behind him.

"No point going away from home if you don't leave your sensibilities behind," he said to himself.

CHAPTER 12

With the help of the excellent mechanical resources of the Evelco factory, Bíró pressed on with his experiments, producing an increasingly promising line of samples of the pen. At the outset he was happy to accept half the wage of the average worker, but after two months he tried to get a raise. He did not get an immediate response from Kovács, who asked him to visit him in his office.

"My dear friend, you must realize that it is not Mr. Goy you're talking to now. We can't afford to pour millions into nothing."

Bíró knew that he would not do himself any favors by entering into a debate, but could not stop himself.

"What do you mean by that, Mr. Kovács?"

"I didn't want us to have this conversation, but if you insist, we will. We have detailed information at our disposal that shows quite clearly that you have not really invented anything at all. Let us take the individual elements one by one. The current design of the tip of the pen is the work of Mr. Goy; it was us who solved the problems with the ink cartridge. And then there is the idea of using an ink paste. . . ."

Kovács took out a photocopy.

"What I have in my hands is a facsimile of a patent registration document. It was made public in Switzerland on May 2, 1938. It reads: "A writing instrument, invented by Wenczel Klimes and Paul Eisner, residents of Prague. Date of registration: April 22, 1936. Parameters: writing instrument equipped with a ball, filled with dye paste." I repeat: it uses dye paste, not ink. So even this idea is not yours. Not to mention the fact that the Czechs had the idea two years ago. Any objections?"

"This isn't the first time I've had this argument. Mr. Goy made similar criticisms. The only thing I have to say is what I said to him then: can you show me a ballpoint pen that works reliably?"

"That is just a question of detail. The most important thing is finally to get our hands on the correct dye—there has to be a chemist who can get it right. Once we have that, all we need to do is put the pen together, and large-scale production can begin."

"I'm not trying to brag, believe me, when I say that of all those who have tried I am the one closest to success. And it is only success that matters, for there is no second place in this race. If it turns out that anyone has beaten me to it, I will pull out immediately and trouble you no more."

"In that case I would be the first to agree. But to return to the question at hand, I cannot raise your salary, I'm afraid. The best I can say is that you stop doing overtime. Only stay in the factory for your regular working hours."

When Bíró finished his day's work he returned to his lodging in Vincennes, a one-room flat near the factory. He wrote a letter to his family, then tried to read, but could not concentrate, and decided to take a local train into Paris.

The city was already making preparations for war. Air-raid exercises were held on a regular basis. Street lighting would be switched off and cars would proceed along the icy streets only by the glare of their headlights. Policemen in phosphorescent white jackets directed the traffic.

Bíró wandered around the city with a map and flashlight in his hand. Only the outline of the Arc de Triomphe, the Opera, and other landmarks could be seen. Near public buildings passers-by would often be asked to produce their documents.

"Vos papiers!" Although the guards would always find Bíró's passport to be in order, their gestures as they handed it back were distinctly uncertain. That evening, around Montparnasse, he saw the figures of

two men in the half-light. Dressed in trench coats, they had cigarettes hanging from their mouths. Bíró took them to be security officials. Though he had no wish to cross their path, neither did he want to draw attention to himself by turning back. On approaching them he was more than a little surprised to hear them speaking Hungarian:

"So, Jancsi, where do you think the Tisza River's deepest point is?"

"Well, we know that the bed of a river gets deeper and deeper, or else the water would turn around. So the deepest point is where the Tisza enters the Danube, at Titel."

"I would say somewhere along its upper stretch, where it flows faster."

Bíró stepped up to the participants in this debate about Hungary's Tisza River and pointed his flashlight in their faces. He immediately recognized his old colleague from Budapest, László Lakatos. The other man, tall with dark hair, he had never met before.

"Do you fellows have nothing better to talk about than the Tisza? Why aren't you talking about French women?"

"Good evening, Edison! I would be grateful if you didn't blind us with that torch!" said Lakatos matter-of-factly, as if they had last met only a day or two before. "You'll soon find out that you have nothing better to talk about, either."

Lakatos gestured to his interlocutor.

"Let me introduce Mr. György János Meyne. He would have a thing or two to say about women, as it happens. As a lieutenant in the Sopron Hussars he seduced his commander's wife. Later he was employed by a German countess as, how should I put it, a trustee of her treasures. According to my latest information he is currently associated with a French milliner lady. Wherever he goes, his women make sure to reward his services rendered."

Meyne laughed to himself quietly.

"Mr. Lakatos is too kind. My own view is that if someone wants a Hussar officer, they should be prepared to pay for one."

"When did you get here?" Lakatos asked Bíró.

"Two months ago."

"It's a shame we didn't run into one another—I was down in

Nice." He gave the newcomer a hug. "Come on, I'll invite you to dinner."

"I don't think you have the money to spend on strangers."

"I am a millionaire without millions, I'll admit, but I'll always have enough francs in my pocket to invite a friend to dinner."

As they walked on, Lakatos continued to philosophize.

"What is it that makes money different from everything else? Most obviously, that money can always be replaced with more money. Other things cannot always be substituted with the same thing. Our attitude to money is also interesting. The majority of people have to struggle to earn it, but find it only too easy to spend it. As if by spending it they can get rid of the bitterness with which it was earned."

They went to the Café des Deux Pistolets for dinner, which was still mostly full of Hungarian émigrés.

"What will be their fate?" asked Bíró.

"They will vanish. German soldiers will be here by next year, asking the band to play the Horst Wessel. And I'll tell you what, they'll play it, too."

"Is it so certain that the French will lose?"

"The French are good raconteurs, but bad soldiers. They have only ever won wars that the Germans have lost."

"Where will you go from here?" Bíró asked Lakatos and Meyne.

"Meyne will be all right; he always finds his feet. As long as there are two slices of bread somewhere, one and a half slices will always be his. As for me, I'm not going anywhere."

"But you are a journalist of international repute. Surely there must be a country that will accept you."

"Well, those countries aren't exactly squabbling over me right now. Anyway, let us say there was one where a desk and good money were waiting for me and where, instead of cheap department stores, I could permit myself to frequent those elegant little shops where they only have a clothes hanger or an antique chest in the window. Where I would have an apartment with my Voltaire books, medicine, and corkscrew all in their proper place. That, I think, would be the worst thing in the world."

"Why?"

"Because, out of gratitude alone, I would become an ultraconservative servant, a servant to the regime I would have to thank for my safety and well-being. You know that is not for me. Neither would it be a stable state of affairs. There would come a day when things would start going badly there, too. The ground would be cut from under my feet again, and I would have to go back to the atlas and look for another new home."

"But you managed to leave Hungary."

"It's not as simple as that. It wasn't me that deserted Hungary, it was Hungary that deserted me."

They were joined at their table by a lean, elegant, gray-haired man wearing the ribbon of the Legion of Honor in his button-hole. He greeted Lakatos and Meyne before introducing himself to Bíró.

"My name is Michel-Marie Bouchonnet. I don't think we've met."

"I arrived not so long ago."

"How are you enjoying these dark Parisian nights? It must be hard to believe that this was once the city of light!"

"I am nonetheless in awe of the people here. If they were not so careful, half the population would have been run over by the cars in this darkness!"

Bouchonnet relished Bíró's compliment.

"In France there is chaos on the streets, but order in people's heads—and that's the more important. Can I ask what you do?"

It was Lakatos who replied on Bíró's behalf.

"He's an inventor. Hungary's Edison."

"Fabulous. I would be very glad to chat with you. Unfortunately, I'm expected elsewhere. It is possible we might be of use to each other," said Bouchonnet, handing Bíró his business card, before stepping out into the street and opening his umbrella.

"Who was that?" asked Bíró.

"He calls himself a retired city councilman, but in fact his influence reaches much further than that. They say he is a member of the internal security office, the *Deuxième Bureau*. I would advise caution."

"Oh, I'm not afraid of the likes of him! He's only the *Deuxième*

Bureau, I'm the *Premier Bíró*!"

"You can already do puns in French! Bravo!"

A few days later Bouchonnet made a surprise visit to Bíró's apartment in Vincennes. The inventor had no idea how he'd gotten the address. The Frenchman took a look around the impoverished room.

"You deserve better surroundings than these."

"I can't afford anything more special, I'm afraid."

"We must do something about that. Could we adjourn to a good restaurant? A fine dinner never does any harm."

Once they were comfortably seated at their table, Bouchonnet came to the point.

"Many people have told me very good things about you, Mr. Bíró. I would like to help you to have the career you deserve here in France. My suggestion is that we form a company, together with my banker friend called Norman, with five hundred thousand francs as initial capital. One of the company's objectives would be to develop, patent and sell your invention."

"And what would my input be?"

"Your work and your ballpoint patent. The direction of day-to-day activities would be in your hands; you can employ as many people as you think necessary."

"But I have existing obligations to another company."

Bouchonnet smiled.

"I can give you a guarantee that the firm in question will not register any objection. The decision is yours, entirely yours. I have drafted a preliminary contract: please have a look at it."

Bíró was wise enough by now to realize that this contract tied his hand completely. Yet he had no choice but to accept it. Bouchonnet was offering him many times what he was getting paid by Evelco, thus offering Bíró the chance to support his family back home, maybe even to pay for them to come after him. He made only one condition, as a form of "escape clause": that he be able nominally to keep his job at the Evelco factory.

Once the contract was signed, Bouchonnet arranged for Bíró's

French residence permit, now close to expiring, to be extended by three months.

"Why won't they grant me a permanent resident permit?" asked Bíró.

"It's still a little early for that. All good things come to those who wait," smiled Bouchonnet in reply.

The Société Française d'Applications des Brevets L. J. Biro was formed. Its owners rented a large apartment, using two rooms for offices and the third to store the necessary equipment; the experiments themselves took place in the kitchen. The only other employee was a Hungarian-born artillery captain by the name of Madarász, who spoke reasonable French and had a certain amount of technical experience. To begin with, Bíró could not understand why he had accepted this badly paid job, then realized Madarász had caught himself in a bureaucratic trap. He had spent a long time in prison, and it was only thanks to Bouchonnet's intervention that he had not been thrown out of France on being released. Madarász had no flat of his own, and Bíró was obliged to turn a blind eye to his sleeping in the office on a handful of cushions on the floor.

Finding the right dye paste continued to prove the greatest problem. Though every wall of the kitchen-turned-laboratory was covered with stains, the experiments still refused to provide results.

György Bíró was simultaneously working on the problem in Budapest. They would inform each other of their progress, even sending each other samples, but to no avail.

"We were beginning to think we had a curse on us," László Bíró was to recall, "and that the composition of the dye we were exchanging was being transformed as it crossed the Hungarian border. We could only assume that the quality of the materials was radically different in the two countries, or that it was due to the difference in climate. One thing was for sure: we still couldn't find the answer we needed."

CHAPTER 13

When Imre Gellért arrived in Paris on March 9, 1940, he was the bearer of bad news. Bíró's wife Elza was working as a seamstress in order to support herself and her daughter, but had to undergo an operation for her apical lesion, and while recuperating was unable to work. The whole family was dependent on the elder Bíró's dentist's salary and the money sent from Paris.

Elza had not fully recovered when the police came to her house and took her to the financial authorities. Acting on the assumption that Bíró had sold his patents abroad, they accused him of transferring his foreign currency earnings home illegally instead of sending them through the National Bank of Hungary. They demanded that Elza reveal where she had hidden the money. After the unsuccessful interrogation came to an end, they let her go, but threatened to summon her back soon.

"We have to do something," said Gellért. "It wouldn't be unlike them to lock her up in prison, and her lungs will get inflamed again."

"You're right," Bíró replied. "I would have liked to wait a little longer, but that won't be possible now. When you go home, tell Elza to sell everything, sort out her passport, and come with the kid to Paris."

"What will you have to live on?"

"I have a friend whose fiancee runs a hat manufacturer. I'll ask her to give Elza a job."

As they went to the café to have dinner, their mood was sour. Bíró spoke first.

"How was your journey?"

"I went to the Argentinean embassy in Budapest, but the train ticket wasn't there, and they didn't know anything about the hotel reservation, either. I had to ask Andor Goy to cover my travel expenses."

"How did that go?"

"He hummed and hawed a bit to begin with, saying he'd already paid us too much, and what guarantee did he have he'd get the money back? I said I wasn't about to leave my mother in the lurch, and that I'd settle the debt on my return. I want to sell off the travel agency, anyway; I should get enough for the furniture and the two typewriters to cover what I owe him." Gellért put down his fork. "Laci, what's going to become of this? We've hardly begun and the Argentines have already reneged on their promises."

"Maybe this is just the famous 'mañana, mañana.' We'll soon find out."

Bíró and Gellért went to the Argentine embassy at ten o'clock the following morning. On their presentation of General Justo's business card, the porter directly led them up to the commercial secretary's office.

"I do apologize, but the postal service returned the train tickets and the travel expenses with a note to the effect that Mr. Bíró had moved to an unknown address. What do you think is a reasonable amount to cover your costs of coming to Paris?" asked the official, and gave them the money immediately. He then took them to the meeting room, where a committee of Argentine businessmen was waiting for them. Sitting at the head of the table was Justo, looking rather tired. Once he had dispensed with formalities he handed over the chair to Béla Bernáth, a Hungarian-born plastics manufacturer from Buenos Aires, who was also acting as the interpreter.

There was only one topic of discussion: the ballpoint pen. Bíró gave his introduction to the invention in his usual way, explaining that while the current model was useable, it was not a definitive version. He promised that they would soon create a prototype suitable for mass production. Bernáth interrupted him.

"Why did you come to France to continue your experiments?"

"This is where I currently have the best working environment.

The idea is not tied to any particular country, however; I would be happy to discuss the possibility of production and sale in Argentina." Bíró placed some samples on the table. "Try these out for yourselves, gentlemen."

The Argentines kindly asked the two Hungarians to adjourn to the embassy cafeteria. As they sipped their coffees, Bíró and Gellért could hear the noise emerging from the meeting room. From the raised voices they could tell that the discussion was animated. Gellért sighed.

"This is when it would be good to know some Spanish."

"Do you think we have a chance?"

"So long as they have something to discuss, anything is possible. Could you imagine living in Argentina?"

"In the words of the great László Lakatos: 'I would live in Hungary, if only they'd let me.'"

A full half-hour passed before the two were invited back into the room. As Bíró took his seat at the head of the table, he could not help noticing that one of the Argentine negotiators had got his hands dirty with ink. Bíró whispered to Gellért:

"We have about as much chance as a rocking horse at the Kentucky Derby."

Bíró tried to make eye contact with Justo, who made no reply other than a sympathetic gesture with his hands.

"We have tried to examine the invention from every possible perspective,' began Bernáth, spokesman for the group of businessmen. "And we have decided that it would be best to wait until you arrive at a definitive solution. The moment you do we will leap into action, determine ownership of the product, and most likely form a joint-stock company that would register the patent in Argentina—at its own expense, of course. We would order your plane tickets, arrange residence and work permits, and cover all the costs of your stay. We hope it will not be long before you come up with the perfect ballpoint pen and we are able to welcome you to Argentina! Until next we meet, Mr. Bíró!"

General Justo also said his goodbyes.

"I'm sorry I couldn't do more for you. If you manage to find some other source of financing, you can always get an Argentine visa by showing my business card. Bonne chance!" By means of consolation he offered Bíró a cigar.

They stepped out onto the street. Bíró stopped, and leaned on the nearest wall.

"I am fed up with always falling at the final hurdle!"

"What options do you have?" asked Gellért.

"I have just one joker left among my cards, the contract I signed with Mária Pogány. I'll write to her to say that I am happy to meet her conditions, and ask her to send me a boat ticket and cover the costs of my stay in Argentina."

"Don't take this the wrong way, old man, but that's a pretty long shot."

"No risk, no reward, as they say." Bíró suddenly held his hand to his forehead. "Oh God, I haven't gone and left Maria's card at home in Budapest, have I? Let's hurry, before I go mad."

They hailed a taxi. At home, Bíró rummaged through his papers and emptied the drawers of his desk, but still he did not find the business card.

"I can't find it anywhere."

"Sometimes it's not a question of when you find it, but when it finds you. Just relax, and start again from the beginning."

Bíró shook everything out of his suitcase, and when the card dropped out of one of the pockets in the lining, he picked it up and kissed it. He sat down and as a reward for his efforts he lit the cigar Justo had given him.

"You must admit I deserve this much, Imre."

The following morning they went to the post office to send a telegram to Mária Pogány. Bíró mulled over what he should write.

"I don't want to set off for Argentina alone. I'll ask her for two tickets, saying that I cannot work without a partner."

"Who would that be?"

"You, if only you'd change your mind."

"Don't get your hopes up!"

"Meyne, then, he's an old hand, he'll do well out there."

"He's really going to leave them all in the lurch, waiting at the church?"

"Why the hell not? There are rich widows in Argentina just like anywhere else."

As Bíró finally handed over the completed telegram form, it was as if all his remaining energy had left him. He stretched out his arms, exhausted.

"That's my last punch. If this one doesn't land, I'll throw in the towel."

"Don't say things like that, Laci. You know what they say: never say never."

"Not bad, but a bit past its sell-by date."

CHAPTER 14

B ack in Hungary, Gellért tried to make arrangements for Bíró's wife and daughter to travel to France. The task was not an easy one, and it was only at the end of July that they reached Paris. The family's first concern was to establish its day-to-day routine: Elza began work at Meyne's hat-making facility, and Mariann was enrolled at a children's camp to help her learn French as quickly as possible.

In his spare time Bíró would take his family out to see the city. Paris was enjoying its last days of freedom; as was to be expected at such times, the theaters, cinemas and night spots were crowded. It was not easy to find a seat on the café terraces, either. The café folk discussed the supposedly invincible Maginot line, never forgetting to mention the name of the legendary General Weygand, who had only just been brought back from retirement and placed at the helm of the French forces.

The impending crisis could easily be felt within industry and commerce. Bíró witnessed the changes in priorities with his own eyes: Evelco scaled down the preparations for production of the ballpoint pen, then stopped them all together, as the factory was put to military use.

From then on Bíró's livelihood depended entirely on Bouchonnet, who strove to exploit this situation as best he could. As Bíró later describes in a letter:

> With nothing by way of explanation, he asked me to transfer all my rights concerning the ballpoint pen to the "Société Biro." Meanwhile he failed to

transfer the money to the company that already formally appeared on the books. He didn't pay my salary either, without which I was unable to support myself.

One day an argument erupted between us about this matter. In the presence of Madarász, the man in our employ, he stated that:

1. If I were unwilling to relinquish my rights to the ballpoint pen in France, he would have me sent to a concentration camp within twenty-four hours—something he, as a councilman in the Saint Clou district, could easily do to a Hungarian citizen.

2. If I were thinking about leaving the country, I could forget about it, because he had made arrangements with the police barring me from receiving an exit visa.

Madarász tried coming to Bíró's defence, only to be threatened by Bouchonnet himself.

"If you don't keep your mouth shut, I will not help you to extend your residence permit. Your type, the *'sale étranger,'* are kicked out of the country in twenty-four hours."

The former artillery officer was enraged, and threw himself at Bouchonnet, and had Bíró not intervened and pulled him back, he would have strangled him. As it was, he tore the skin on the Frenchman's neck; blood stains trickled down his crisp white shirt. Madarász continued to hurl abuse at Bouchonnet as he left, promising to come back and finish the job.

It seems that this little incident had a good effect on Bouchonnet; he apologized to Bíró, and even paid him some of what he owed him. Work on the pen came to a halt, with the Frenchman promising to make contact again when the time was right.

The inventor continued to have little reason to be optimistic. All good things come to those who wait: the looming crisis finally came to a head. On September 1 the Germans attacked Danzig; two days later France declared war on Germany, and World War II was in full force.

Fighting had not yet begun on the front, but the German offensive was expected at any moment. In Paris, air-raid sirens would often sound—for the time being only for test purposes. People would grab suitcases containing their most important belongings and head for the nearest metro station.

The French would go to drastic lengths to suppress their fear, and slogans like this one were regularly heard on the streets:

"If we have to, we'll set out to war in a taxi again, just like in 1914, when the Boches were threatening Paris."

Xenophobia was increasingly rife. Anyone heard speaking a foreign language in a public place was set upon.

"*En français, comme tout le monde!*" the crowd would exclaim, and the perpetrator would thank his lucky stars if this was as far as the insults got.

In these circumstances Bíró decided that the lesser evil would be to send his wife and daughter back to Hungary. After only a month together they were separated again. And now they had even less hope of being reunited than they had had on New Year's Eve of the previous year.

The inventor had no idea what turn his life could now possibly take. His last hope, the telegram he had sent to Argentina, generated no response. In the absence of other options, he spent whole days sitting in the Café des Deux Pistolets with the other Hungarians of Paris.

The number of his fellow countrymen to be found in this city dwindled continuously. Some enrolled in the French army, others were arrested by detectives in one of many raids. Lakatos was one of the few friends who remained. Though only sixty-two, his health problems made him look like an old man; even in the heated café he sat with his scarf wrapped around his neck.

"There's nothing worse than when a man is at odds with his own body," he said, pointing to his chest, "and this is the ungrateful body I have had to carry with me across every frontier all these years."

By the middle of September, Lakatos had also resolved to interfere in his destiny; he left a note with the waiter telling Bíró that he had a ride offer to Nice.

"I have set off on my trek into the void, a place I'll get to the minute I depart. See you at Philippi."

Bíró's other contact, Bouchonnet, had not been in touch since war had broken out. Bíró had written him off completely when, one evening, he made an unexpected appearance at the café.

"Good evening, my friend. I'm so pleased I've found you. I desperately need to speak with you—but not here. Let's go to Boulevard Haussmann."

"But we gave up the lease on that place!"

"I took it out again."

When they were in the apartment, Bouchonnet cautiously closed the door and sat down next to Bíró.

"I didn't want to talk to you about this in the café: it's full to the brim with secret agents and informers."

"Even among the Hungarians?"

"Oh yes. Don't take this the wrong way, but some of your compatriots are only too willing to get involved in certain undercover tasks. Especially the journalists."

"You know that from first-hand experience?"

Bouchonnet did not blink.

"No, from a good friend. And the issue in question is particularly confidential." Even in the locked apartment he kept his voice to a whisper. "In actual fact what I want is to pass an offer on to you. Certain official circles would be willing to extend your stay in Paris—indefinitely. They could pay for your living expenses, too, if you would be willing to apply your skills to the French war effort."

Bíró was so taken aback he did not even think of asking Bouchonnet how he came to be in touch with these "official circles" and how those circles had come to know of his existence.

"What do they have in mind?"

"In our previous conversations you mentioned that you have some ideas for innovations in military equipment. These official circles would be interested in even the smallest of those ideas. Sleep on it. We'll meet again tomorrow morning, here, and discuss the details. It goes without saying that this matter is top secret, that you must not mention it to anyone."

Bíró found it hard to believe that one of Europe's largest fighting forces should really be in need of his assistance in its war effort, but he had long stopped being surprised. He knew only too well that he was in no position to refuse the request, if he didn't want to end up in an internment camp. He went home and looked right through his ideas dossier, picking out a number he thought might be of interest. He took them with him the next day. Bouchonnet did not arrive alone. He did not introduce the other man by name; all he was willing to reveal was that he was a high-ranking official at the ministry of defense. Bíró showed them his various ideas in turn. The newcomer displayed an exceptional familiarity with military technology. He showed an interest in all of Bíró's suggestions, but selected one in particular as being the most significant. This was a design for a flammable chemical compound that could not be extinguished by water— a compound that, dropped from a plane, would be capable of setting huge areas alight in a very short space of time. It was, in a sense, a predecessor to the napalm bomb.

The official asked Bíró to begin work on the project immediately and keep him posted on the progress of his experiments. To all intents and purposes the inventor moved into the old kitchen-cum-laboratory on a permanent basis. With Bouchonnet's permission, he took on a partner to assist him in the form of Ferenc Ruzsák, a Hungarian colleague from the Evelco factory.

Bíró describes the essence of the invention as follows: "I mixed potassium chlorate with carbon tetrachloride and white phosphorus in the right measure, and sucked it up with wood-pulp. This mixture keeps indefinitely if locked in a vacuum, but if exposed to air, the carbon tetrachloride soon evaporates. After it does so, the phosphorus heats up, then ignites, and blows up the potassium chlorate. Sealed in a thin-walled glass container, it would be suitable for use as an incendiary bomb to be dropped in large quantities from an airplane."

Bíró and Ruzsák wanted to be able to present real results as soon as possible; they worked day and night, taking shifts in order to get just a few hours' sleep. One day Bíró was awoken by a huge explosion, followed by the sound of crying. Ruzsák had fallen asleep with his head on the apparatus. One of the containers had exploded from the pressure, and its contents had sprayed in his face. Bíró rang for an ambulance at once, and accompanied Ruzsák to the hospital. He returned to find a policeman guarding the house; Bíró readied himself for arrest. What could possibly seem more suspicious, he asked himself, than a nervous foreigner secretly working with explosive materials in wartime Paris, and injuring someone with them? Bouchonnet was soon on the scene. Bíró told him what had happened, but the Frenchman dismissed his worries with a wave of his hand.

"I know everything. The case is closed—I'll send the policeman home, you get on with your work."

But Bíró could not bring himself to touch the apparatus that evening. He sat in his chair for hours, his head in his hands. He knew that he should be glad to have got away so lightly, but he could not help realizing his servitude. So long as Paris was still in French hands, Bouchonnet and his superiors had him at their beck and call; if the Germans were to enter the city and find this "non-Aryan" foreigner making weapons for the French army, there was little question what fate would befall him.

In the end he decided to go home to his lodging. He took a sleeping pill, but this did not stop him from staring for hours at the pattern on the ceiling.

CHAPTER 15

Nothing stirred in the apartment—the principal tenant, a divorced, middle-aged man, had already been conscripted— and Bíró only woke up around noon. He hurried to get dressed and set off for work. He was about to close the door of the building behind him when his eyes came to rest on his mailbox, which he had not looked at for weeks. It was overflowing with papers. On opening it, Bíró found a letter addressed to him among the many wartime pamphlets. The sender was marked on the envelope: Mária Pogány, Buenos Aires. There were also three letters from the French National Bank, each stating that a thousand dollars had arrived in his name, and could be collected during office hours. The last notice contained a deadline: were the money not to be collected within forty-eight hours, he would be considered to have been mobilized, and the money returned to its sender.

Bíró could not make out from the smudged date stamp on the envelope whether the deadline had passed or not. Even in the cold of winter Bíró broke out in a sweat at the idea that he might be too late. He looked for a taxi, but in Vincennes they had all been put to use by the army or by civilians fleeing to the south, and so he had no choice but to take the *métro*.

Crouching in the corner of the carriage, Bíró opened Pogány's letter. In it she began by describing her wedding, and how well she was getting on with Lajos Láng, her new husband. She went on to say that she accepted his offer, and would be glad to welcome Bíró and his colleague in Argentina. She had transferred the money to Paris at the same time as sending the letter, and arranged for the boat tickets to be waiting for them in Barcelona.

Bíró made it to the bank before it closed. His pessimism proved to be justified. The thousand-dollar draft had already been filed for return, and it took lengthy persuasion for the reluctant clerk to cash it.

It was only after he had completed this task that Bíró suffered an attack of nerves; weak at the knees, he had to sit down. His thoughts turned to his next move. The most important thing, he considered, was to avoid Bouchonnet suspecting anything. He was, after all, in a position to use his connections to prevent Bíró leaving France. He decided to go to the laboratory and start work. Bouchonnet was already there waiting for him, helping to clean away the debris from the previous day's explosion.

"Where have you been, my friend?"

Trying to insinuate that a racy escapade was behind his tardiness, he shrugged his shoulders.

"She was quite the passionate type—for a man my age, anyway."

Bouchonnet patted him on the back.

"Congratulations! I am glad to hear you are beginning to enjoy things here in Paris. I visited Mr. Ruzsák in the hospital this morning and his eyes have not gotten better yet, I'm afraid. I'll send you a replacement assistant."

There was no doubt in Bíró's mind that he would send an agent of the secret police to take up the role, tightening the control another notch. If he wanted to do something, he had to do it quickly. He would have liked to take a few samples of the ballpoint pen with him, but Bouchonnet had all of them.

It was late in the evening when Bíró arrived at Meyne's apartment. Fond as he was of the former Hussar, he did not trust him with all of the money, and gave him only five hundred of the dollars, as well as Justo's business card.

"Show this at the Argentine embassy and they will grant us visas. Then go and buy train tickets to Barcelona."

"I don't have any time tomorrow; I'll do it the day after."

"Put everything else aside. It is only a question of time before the French-Spanish border is closed, and then the mousetrap will have us surrounded once and for all. Goebbels has promised to take Paris by the fifteenth of next month."

The worldly-wise Meyne wasted no time. In two days he had sorted out all the necessary papers; Bíró packed together all his things the same evening. He could not rule out the possibility that his house was being watched by agents of the Deuxième Bureau. Taking his suitcase would have given away his intention to travel, and so he packed his personal belongings and his ideas dossier into his larger attaché case.

Although the train departed at noon, Bíró could not stand the confines of his flat any longer, and left for the center of Paris with the first *métro*. He wandered the streets at a slow pace. Huge posters were being posted on the walls of downtown buildings: "*Confiance! Sang-froid! Courage! Weygand vaincra!*" Elsewhere a map of the world had been put up, with the allied empires of Britain and France in red and Germany in gray, with the caption: "We will win, because we are stronger!"

The scene on the streets ran contrary to these slogans. Doors of shops were in many places covered with signs saying, "Closed due to mobilization"; columns of trucks were making for the Porte d'Orléans in a southerly direction; from the chimneys of the foreign ministry at the Quai d'Orsay, a thin line of gray smoke was winding its way into the sky.

"They're already burning their documents!" Bíró muttered to himself.

He had agreed to meet Meyne under the great clock at the Gare de Lyon. As he climbed the steps leading out of the *métro*, Bíró was struck dumb by what he saw. The train station was closed, and policemen in white jackets blocked the entrance, but all the people gathered about hoped they would eventually be allowed to board their trains; they refused to leave, lying on the ramp or next to the walls with their heads on their luggage. Wave upon wave of new people were joining them from the *métro*.

There was only an hour left until their train was due to depart, and the police cordon seemed impenetrable. Bíró was well aware that the best option in the circumstances was to consider the way in which he could continue his life in Paris, but he was not willing to ruminate on this. He was sure that he would not fall at the final hurdle, that he would still be able to escape from France.

Meyne arrived much later than agreed. Bíró was overflowing with anxiety, but tried to control himself.

"Looks like we've missed it, then."

"No, it's still there on track four. The only question is whether we can get on it or not."

"And what decides that?"

"Have you got a hundred dollars for me to bribe the police?"

Bíró handed him the money.

"Sure that will be enough?"

"Easily. I could buy the interior minister himself for two hundred! Move back a bit from the crowd."

Bíró took his briefcase and stood in a phone booth. To avoid being disturbed, he picked up the handset and pretended to conduct a conversation. He got so caught up in his thoughts that he imagined his daughter was at the other end of the line, and continued to talk as if she were really speaking to him from all those hundreds of miles away. His nerves were in tatters from the tension; he had difficulty choking back tears.

"Put this on!"

He looked up. Meyne was standing in front of the phone booth, trying to hand him a white police jacket and hat. He was already wearing his. They set off, the crowd made way for them and when they reached the top of the ramp the barrier was lifted for a brief moment. Once inside, they returned the uniforms, Meyne got his suitcase back, and they ran for track four.

There were only a handful of passengers on the train, and the two Hungarians had no trouble finding a compartment to themselves. Bíró could not help being reminded of his departure from Budapest on New Year's Eve of the previous year. What if Buenos Aires would be just another temporary stopping point in his life, and he would have to repeat this departure a countless number of times? Taking out his passport, he put his briefcase up on the luggage rack.

"Is that all your belongings?" Meyne asked.

"In an ideal world, I would pin my toothbrush in my buttonhole and leave it at that. I hate carrying things about."

The train made long stops along the way, sometimes taking detours, including one of hundreds of miles through Nîmes and Montpellier; it arrived at the border station of Port Bou only the following day. French customs wasn't operating. Word spread that on the previous day, May 12, the "phony war" had come to an end. The Germans had attacked France, armored units had crossed the Ardennes, taking the Allied forces from behind and advancing toward the sea. The whole state apparatus had collapsed under the weight of the catastrophe.

In the ensuing chaos no one wanted to be responsible for deciding whether the train could continue its journey, and it was hours before it was shunted over to the Spanish side of the border at Gerona. Bíró's small amount of luggage aroused suspicion. The Spanish customs guards took him to one side and undertook a painstaking examination of his briefcase. The head of the customs station looked through the papers in his dossier one by one. When he came to the cross-sectional blueprint for the ballpoint pen, his face paled.

"This is the design for a torpedo! How did this come into your possession, and who do you aim to pass it on to?"

Bíró's flimsy French skills were not up to convincing the official that the drawings depicted not a weapon but a new type of writing instrument. In the end he showed him the facsimiles of the patents for the pen. The customs guard's mind was put at rest, and he stamped Bíró's entry visa obligingly.

Upon arrival in Barcelona, the two Hungarians would have most liked to continue their journey at once, but their boat tickets had not yet arrived. They had no problem making ends meet, however; for Mrs. Láng had booked them rooms at the Hotel Continental. The Continental was one of the best hotels in the city, but this did not stop

Bíró feeling uncomfortably trapped.

"Every place is a prison if you cannot leave it whenever you like, in whatever direction you please."

The Spanish Civil War had come to an end only months before. The debris was still being tidied away, and the country's economy was in tatters. Even at the luxury hotels, meat was served only twice a week, the day after bullfights. Most people had difficulty finding provisions, and gangs of children roamed the streets looking for food in trash cans.

As he had left France with no warning, Bíró was reminded of the fact that his family and business associates would be worried about his well-being. He wrote letters to his wife, to Andor Goy, and to the owners of the Evelco Company. He also decided that he owed Bouchonnet at least a formal explanation. He informed them all of his present situation and of his hopes regarding his future in Argentina.

When the ink had dried on the last of these letters, Bíró was about to set off to look for a post office that was open, but Meyne held him back. He warned him against sending them immediately.

"Wait until we are just about to leave, and then post them. It's better if no one knows that we are in Barcelona at the moment."

"What do we have to be afraid of?"

"The city is teeming with foreign spies. You know how many émigrés they've murdered already."

Bíró shook his head.

"In my experience, as you flee your fear grows. In any case, there's no great price on my head; an attempt on my life would be an honor. So, if you don't mind, I'll head off for the post office now."

Bíró received replies to all these letters while still in Barcelona. Unsurprisingly, the nature of the responses varied strikingly. Elza reassured her husband that he had made the right choice, that this was the only way the family would have the chance to meet again. She had already begun to learn Spanish, and added: "*Hasta la vista en Argentina!*" Kovács, director of Evelco, wished Bíró the best of luck, asking him to send a report on the opportunities offered by the Latin

American market when he had the opportunity to do so. Bouchonnet accused him of ingratitude in deserting a country that had accepted him and had put a roof over his head, and of lacking faith in its military strength. He would only be willing to forgive Bíró, he wrote, were he to relinquish all his rights to the invention in France.

Bíró described the fourth reply in his memoirs as follows:

"Andor Goy's letter was full of crass insults and threats. According to him, my 'escape' was a typical example of cowardice, further evidence that I was a 'traitor'. My escape was completely in vain, as however far I got from them, there would be no place on Earth where they (he had written 'we') would not reach me. He made mention of the superiority of the Aryan race."*

After the replies arrived, there was not much more mail to speak of, and the days passed in such a monotonous fashion that Bíró thought they might as well have been carbon copies of each other. He and Meyne had long run out of subjects of conversation, and Bíró preferred to spend his days sitting on a bench by the sea. He would take a bottle of orange juice with him, and as he sipped away he would watch the ships sailing away on the ocean as far as the eye could see. It was there that Meyne found him on one afternoon.

"You must come at once!" he gasped, panting.

"What's happened?"

"I got a call from the travel agency. Turns out there's a double cabin available. They'll let us on if we can get ready in an hour."

They ran off down the road toward the hotel. Later, Bíró would often be reminded of the half-full bottle of orange juice he had left on the bench. This became his last memory of the taste of Europe.

The two Hungarians made it to the ship, which was in the process of making its departure. The Cuidad de Sevilla, a 6,000-ton steamer,

---

*This letter would indeed put Goy in a bad light. I use the conditional on purpose, however; for there is no evidence that he did in fact write it. Thirty years later the two former business partners were on two sides of a trial at a Nuremberg court. The matter of this letter came up during the case, but Bíró was unable to produce it as evidence, and so it was not accepted. His claim looked suspiciously like an attempt to discredit Goy, something there was no lack of on the other side, either. Goy writes the following words about Bíró's book *Silent Revolution*: "I could not find one honest sentence in it to accompany the flurry of self-praise. It is arguably unworthy of being put through a printing press."

had formerly been used as a pleasure boat between Barcelona and the Canary Islands. During the Civil War it had remained in the port, only to be discovered by a new shipping company when the war was over. The company had bought it, and in light of the boom in travel generated by World War II, transformed it for transatlantic routes. The crew was bundled together in haste; the captain himself was making the trip across the Atlantic for the first time.

This particular journey had been rented out by a few rich South American families to bring home relatives who had remained in Europe. A number of Rothschilds were on the passenger list. Outsiders could, at best, hope to find space in the remaining cabins, but Bíró and Meyne were lucky enough to be allocated the numbers 171 and 172, and be placed in first class.

Once on board, the inventor instinctively loosened his tie: he felt that he had just made his escape from hell. He and Meyne unpacked their things in their sleeping quarters and made their way to the dining room. They had hardly made themselves comfortable at their table when the gong sounded to indicate that the ship had embarked. Practically every passenger was leaving his or her homeland to avoid Hitler's grasp: on hearing the gong, all stood up and clapped.

The waiters brought enormous portions—the chef had prepared more than a pound of meat for each passenger.

"We Argentines eat more meat than any other nation on Earth," said one of the waiters. "But with forty-five million cows on the pampas plains, we cannot possibly eat enough for it to run out."

After the sparse courses served in Barcelona, Bíró and Meyne ate so much they were hardly capable of following the others into the salon afterward. Without formal prompting, all the passengers had begun to introduce themselves to one another. Hispanic surnames were accompanied by plenty of Italian, German, and Slavic ones. Most of those on board were second- or third-generation Argentinean citizens, as if to prove that World War II was not the first time people had fled persecution in Europe for safety in South America. Bíró thought aloud:

"If such a variety of people have succeeded in fitting in there, why can't we?"

"Time will tell. Did I tell you the story of my uncle Fábián?"

"No, what happened to him?"

"He once made a sea crossing, during which he wasn't allowed to drink beer. He died."

"I get the message, old man," said Bíró, "Let's go up to the bar on deck—my stomach is starting to churn."

"Try to keep looking at one fixed point, like the horizon—that's the best thing."

CHAPTER 16

In August 1940 the Atlantic was still free from hostilities, but the captain's inexperience often brought the Ciudad de Sevilla into danger. On one occasion it barely missed an underwater reef; on another, misinterpreting a meteorological report, they found themselves in a storm.

The passengers were too troubled with seasickness to come out of their cabins very much. The experienced Meyne attempted to alleviate the suffering the voyage was inflicting on his friend, giving him onions to relieve his sickness and telling him to put a cushion under his stomach when sleeping. Bíró was nonetheless much relieved when the ship's band began to play *Aquarela do Brasil*, the traditional signal that they were soon to disembark on the coast of South America.

The two Hungarians stood on deck and watched as the Ciudad de Sevilla docked in the Brazilian port of Santos. Little boats doubling as corner shops maneuvered themselves around the ship. Bíró bought a large bunch of bananas from one of them, then waved his last ten-dollar note in the air.

"That's it, no more after this has gone."

"Pretty tight. When do we get some more?"

"Don't know."

"If everything goes according to plan, we'll arrive in Buenos Aires tomorrow morning. Mrs. Láng will be waiting for us, surely she'll have a peso or two."

"Begging is not the way to kick things off. What matters is that she gives us a little workshop with the right machinery, that we have trained professionals at our disposal, then we get right down to work."

"You mean, me too?"

"Yes, you too."

"Sorry, Laci, but don't count on me. You know what I always say: if you want a Hussar officer, you should be prepared to pay for one. If you keep going on about having to work, I'll swim all the way back to Europe."

Bíró looked at Meyne.

"You don't believe in this whole ballpoint thing, do you?"

"In my honest opinion? No, not really. You've been struggling with it for years now, and there's still something missing."

"I can understand that. It's only human nature to make judgments with the benefit of hindsight. If this ballpoint pen is a success, the Hungarians will say that I am Hungarian, and the Jews will say that I am Jewish. If I make a fool of myself with it, the Hungarians will call me a Jew and the Jews will call me a Hungarian."

"That sounds about right."

"There's something I must tell you: however things work out, I don't want to hold you back. If you find better opportunities elsewhere, you are free to go."

"Thanks. If I do, you'll be the first person I'll tell."

That evening the ship continued its journey, traveling up the Rio de la Plata, and at four o'clock the following afternoon, arrived at the southern Buenos Aires port of Puerto Suro. The river widened into a delta at this point and the almost static water was surrounded by brown reeds, among which wild ducks were swimming around. A few abandoned ships had been towed towards the riverbank; they were engulfed by rust, and all around them rubbish was floating in the water. Flocks of birds swooped down to eat the seed that had been scattered during loading. The thick fumes rose up in clouds, disappearing into the mist some five or six meters above the factory buildings.

With so little baggage it did not take Bíró and Meyne long to get through customs and climb down onto the jetty. The crowd that had assembled to welcome the passengers of the Cuidad de Sevilla soon dispersed, but there was still no sign of Mrs. Láng.

"What do we do now?" Meyne asked. Bíró shrugged his shoulders.

"We do nothing. We'll find somewhere to sit down. If anyone is coming for us, this is the place they'll look."

They set off alongside the streetcar tracks that led to the city; the stench of tar and sewage was stifling. They waited for a yellow dual-axle streetcar to pass by them, then crossed the tracks.

At the edge of a slum they came across a bar cobbled together from corrugated iron, the front of which was made up of the flattened frame of a truck. Bíró took a look inside. Gray sausage had been prepared on a tin tray, ready for cooking; further inside, he saw a shabby refrigerator, around which a wire had been tied to keep the door closed.

"Hello! Is anyone there?"

An old man with short black hair emerged from behind the scenes. His left eye was half-closed: it was hard to tell if he could see with it. Bíró ordered two beers, then they took a seat in the shade of the only tree in front of the bar. The barman served their beers in metal cups. Bíró was careful to clean its rim before taking a sip.

"Cheers!"

"Cheers! Don't look so glum, Laci. Even if this Láng woman has vanished into thin air, all is not lost. You know what they said about Hungarians: they enter a revolving door behind you and come out in front of you."

"Bit passé, but I hope you're right."

Meyne was watching the passing traffic.

"Look how strange the cars are. No windshield wipers! You can imagine how often it rains."

One of the cars traveling from the direction of the city was a huge Packard. It slowed in front of the corrugated iron bar. The man who got out of it was about fifty-five, and wore a European-style, narrow-brimmed hat.

"Mr. Bíró, I presume? I'm Gábor Müller, Mr. Láng's partner.

Mária sent me—I'm sorry to be late." He glanced at Meyne's suitcase and Bíró's bag. "They are unloading your baggage later?"

"No, this is all we have."

"And where is the equipment needed for the production of the ballpoint pen?"

Bíró pointed to his head.

"In here."

"I understand," said Müller, though clearly he didn't. He opened the car door. "Please, get in."

Müller drove his two guests downtown and took them to a room at the aptly named Hotel City.

"This is just for the time being. You will soon be moved to a guest house where they speak Hungarian. Unpack your things and come to our office in one hour. The address is 92 Reconquista; it's not far— 'Láng and Co.' is on the door. I'll leave you a map with the office marked on it so you won't have any problems."

The two Hungarians had no problem finding their way around downtown Buenos Aires. The city planners had based the design of the streets on those of New York, with a square park in the center, each of its sides exactly one kilometer in length. This distance was divided up by connecting streets, with a hundred houses to each resulting block. This was where the famous Rivadavia began—with over ten thousand houses, one of the world's longest avenues. Meyne noticed that the sidewalk often glittered a silver color. When he took a closer look, he saw that caps from beer bottles had been pushed into the hot tarmac by the thousands.

"As my uncle Fábián would say: this is the kind of place where I could live!"

Bíró stopped at a post office and flipped open the telephone book. Meyne was a little surprised.

"Who are you trying to find?"

"I'm looking for General Justo's number; it's not on his card."

"You think he's going to be in the phone book?"

"Why on earth not? In the Stockholm phone book you can find Gustav Adolf under Gustav Adolf, King."

But Bíró was lost in the myriad of foreign-sounding names, and closed the phone book in frustration.

"There must be some central office where they will tell us."

It must have been around eight o'clock when they found the brass plate inscribed with the Láng name. The gate was already closed, and it was Müller who let them in. A cleaning lady in a blue apron was sweeping in the corridor. Bíró approached her.

"Tell me, love, where is the lady of the house?"

The lady looked up, revealing her identity.

"That's me, Laci, the boss and the cleaning lady in one. There's no job around here I'm not afraid to do." She extended her hand to Meyne. "My name is Mrs. Láng. My friends call me Mária."

Mrs. Láng took off her apron, revealing a fashionable two-piece linen dress.

"Come in. What's the rage in Budapest nowadays? Oh, of course, you've come from Paris, not Pest! How are things there?"

Mr. Láng greeted them in his office. As he stood up from his chair, they could see that he had been sitting on a circular pillow with a hole in the middle, the sort used by hemorrhoid sufferers. The businessman had moved to Buenos Aires from Vienna, where he had for a time been the honorary consul for Uruguay. He had made little effort to acclimatize to the leisurely way they did things in South America, and his environment here was a little piece of Austria far from home. He was wearing an exquisite suit and hat, and instead of offering his guests the customary mate tea he gave them cappuccino and tea cakes.

In their business affairs, it seemed that Mrs. Láng was the one who took the initiative. No sooner had they all begun stirring their coffees than she came to the point.

"It's a long time since I saw the ballpoint pen back in your brother's office, when I had my teeth done. I imagine you've come a long way since then. Can you show us what progress you've made?"

Bíró swallowed hard.

"I'm really sorry, Mária, but I don't have anything to show you. We left Paris in such a hurry that we weren't able to bring away anything at all."

"But surely we could have some specimens sent over from France?"

"I don't think so. The war is raging there; no one has time for this kind of thing."

Mária slammed her coffee cup on the table, irritated.

"So, we bought ourselves a nice little pair of white elephants bringing you two here!" It was obvious from her tone of voice that had she not invested money in them already, she would have sent them back to Europe immediately. Bíró tried to dispel her worries.

"Things are not quite as bad as they seem. We have made very significant progress with the experiments. If we have all the necessary equipment at our disposal, we can prepare new specimens very quickly. We didn't come all this way to hang out in the sunshine."

"What does all that necessary equipment mean in dollar terms?"

Bíró did some calculations in his head: he added together the cost of the machines and other appliances, the necessary materials, made an estimate of the labor costs, and added twenty-five percent to the total for good measure. Mr. Láng was stunned by the size of the amount.

"That's a hell of a lot of money."

"I'm not saying it isn't. But the amount will just get bigger by the day. We must start work as soon as we can."

"You are right about that," said Láng, getting up from his seat. "But we'll return to the financial side of things later. First, I want to show you the workshop. Follow me."

They walked through three large rooms, all stuffed with shelves, containers, and large cardboard boxes containing goods deteriorating under the weight of clothes, ornaments, and sweets.

"Nice little collection you've got here, Mr. Láng!" said Meyne, politely.

"You think so? In my opinion there's only one kind of product worth having, and that's a product you've already sold. But here people don't have spending power. If you give an Indian two bananas, he'll stop working, and so not only will he not earn any more but he will also have nothing more to spend."

As his eyes wandered over the sorry sight of the warehouse, Bíró realized why Mr. and Mrs. Láng had brought them over to Argentina. Their last hope before declaring bankruptcy was to make something out of the ballpoint pen. This thought did not trouble him unduly, but Bíró was worried that they would not have the capital needed to fund the necessary experiments. Finally Láng led the two men to a bathroom about sixty square feet in size.

"This is it, gentlemen, I hope it is to your satisfaction. It's not railway carriages we're experimenting on, after all. The pen shouldn't need that much space." Láng handed them a banknote. "For the time being I'll give you a hundred pesos to cover your everyday expenses, that's more than twenty dollars. Make yourselves at home in your new surroundings. I have some more things to attend to in the office—I'll be waiting for you there when you've finished."

Once they had been left to their own devices, Meyne took a look around.

"Whenever you feel like taking a break for a quick bath," he said, "there's nothing to stop you."

"There is no way I can work here" replied Bíró. "It's a wonder they didn't stick us in the toilet."

"Kick up more of a fuss, and you can get fired, to boot."

"I wouldn't bet on it. We've cost them too much as it is. They want to get a return on their investment."

Láng invited them to join him for dinner at his regular haunt, the Alte Wienerin, a little Austrian-style restaurant. A panorama of Vienna hung on the wall, with the house motto underneath it: "We've never enlarged our rooms, and we've never watered down our wine!"

Sitting at the tables were members of the various waves of emigration from catastrophe in Europe, be they survivors of the pogroms in Galicia or those fleeing from the horrors of the world wars. Even if they were not the greatest of friends, the descendants of German aristocrats sat side by side with communist émigrés in relative harmony. Most of them had picked up the relaxed ways of the South Americans, leaving buttons undone and throwing screwed-up napkins to the floor.

"*Mussen Sie gewören!*" said Láng, lifting his glass. "Let's drink to your successful arrival in Buenos Aires!"

But Bíró was not in the mood for a celebration.

"I think we can make this our first and last toast."

Mrs. Láng frowned.

"What do you mean by that?"

"I am not willing to work in a bathroom. It's not a question of pride. It's just that it's not possible to work properly in such an environment."

"We saved you from certain death, spent a ton of money on the both of you, and now you have the cheek to pick holes in your working conditions."

"We are extremely grateful to you, and rest assured, Mária, we will pay you back every penny that you have spent."

"From what?"

"There's no question that we'll find our feet here in Argentina," bluffed Bíró. "General Justo promised to put in a good word for us if we need one."

"You've hardly got off the boat, and you're ready to make a dash for it! What ingratitude!"

Mr. Láng gestured to them to stop quarrelling.

"There's no point making this so confrontational. We have to wake up to the fact that we're all in this together—we'll find a way of working things out."

His wife said nothing, but it was clear from her expression that she would never forgive Bíró for what he had said. The mood of the conversation failed to brighten up, and they soon bid each other farewell.

"In my opinion," said Bíró, "Mr. Láng is a much better sort than his wife."

Meyne sighed.

"It's obvious that you've never had dealings with the police. They're just using the old 'good cop—bad cop' strategy."

"What's that?" asked Bíró.

"They take the suspect to the police station, where a brutal cop gets set on him at once, shouting at him, threatening him, maybe even giving him a slap or two. The next day the interview is continued by a kind policeman, who gives the delinquent a cigarette, promising him all kinds of concessions. The suspect is touched by this, and shows his gratitude by making a full confession. In our case the wife is the bad cop and the husband is the good one. It's all arranged—don't fall for it."

CHAPTER 17

M r. and Mrs. Láng would not let work on the ballpoint pen commence before the question of the ownership of the invention was settled. They visited a notary public, and Bíró signed an affidavit listing his previous agreements with his partners in Europe, the Goy and Kovalszky Company, and Bouchonnet.

As Bíró explains: "In this document we established the fact that if the annual patent fees were not paid, and the patent not refreshed within a year, protection of the invention would be lost. As the war in Europe made both payment of these fees and the commencement of production impossible, it represented a case of force majeure. As a result of this the listed agreements became void."

The license trial held in Germany twenty years later showed that this legal line of argument had a number of weaknesses. There is no doubt, however, that in South America, at least, Bíró retained full ownership of his invention.

It was on the basis of this declaration that Bíró and the Lángs founded Biro South America Ltd., which had a single commercial objective: to sell the patent for the ballpoint pen.

The partners had to establish their profit share. Bíró, expecting to split profits fifty-fifty, was astonished to hear that Mr. and Mrs. Láng had different ideas.

"We have to ensure a return for Dr. Adolfo Mujica, the lawyer, too."

"On what basis?"

"Mr. Mujica is an important figure in politics here. It was he who helped to arrange your residence permits, and we may need more of his assistance in the future. We shouldn't begrudge him a few pesos."

"All right. What about the rest, Mária?"

"We suggest splitting the remaining profit three ways. One third for my husband and our partner, Mr. Müller, for financing the project. Another third would be for me—for the negotiations held in Budapest and the management of affairs here, including procuring capital."

"But that capital is provided by your husband, who has already got his profit share!"

"From a legal point of view that is beside the point. In my opinion I am not being overcompensated. Not to mention the possibility that I may never get my money back at all. The third share would be all yours, Laci."

"I think my brother György and my partner"—he pointed to Meyne— "also deserve a share for the work they have done."

"We are in complete agreement, Laci. You are welcome to give them whatever proportion of your share you see fit. There you are, the contract. I hope we use the ballpoint pen to sign the next one."

Bíró stared at the paper laid in front of him in silence. His emotions were too stirred by the unfairness of the division of profits for him to sign it. The stalemate was obvious to all those present; there was a long silence before Mr. Láng finally spoke up.

"I can understand what you're going through, Laci—you're not used to the tough conditions of business. I would be willing to add one further clause to the contract: that we accept responsibility for bringing your family out to Argentina within a defined period."

Bíró was in little doubt that the couple were using the "good cop–bad cop" trick on him again, but he didn't care. There was no price in the world that he would not have been willing to pay to save his family from certain doom. He nodded, and signed a document that gave Láng and his wife twice as much share of the profit for their financing and mediation as he got for the invention itself and the many years of effort it had brought with it. He was allowed to take an advance of five hundred pesos a month against his later earnings. This was not only enough to cover his and his partner's living expenses, but he had enough left over to send some to Budapest, too.

"I'm sorry I couldn't get more out of them. We are both in the same boat, of course," he told Meyne.

But the former Hussar officer only responded with an Argentine saying he had just learned, one of several he seemed to have at the ready.

"*El zonzo trabaja, el vivo vive!*"

"What does that mean?"

"The crazy work, the smart live!"

They had already moved to the Hungarian guest house, where the *American Hungarian Worker* was delivered regularly. On returning home, Bíró flicked through the latest edition, his eyes stopping at the headline:

Death of a Hungarian Journalist

He read on:

> A room on the third floor of a hotel in Nice. There was a loud knock on the door. "Who's there?" asked the occupant with a choke in his voice. The reply: "*Tür aufmachen! Gestapo!*" A moment later came the sound of a revolver, and a thin, gray-haired man fell to the ground of the sparse hotel room, blood pouring from his temple. He was László Lakatos, Hungarian writer and journalist, an émigré who fled from Horthy's regime. He was buried in a pauper's grave by the cemetery.

Bíró put down the paper.

"There's no escape, not even for the Master!" he muttered to himself.

Bíró set to work with all the energy he could muster, as if this would reduce his wait to be reunited with his family. Láng kept his promise to improve their facilities, and found a suitable location downtown, at 3040 Calle Oro, where Bíró set up his workshop. Sitting at his desk, he gathered together his thoughts and experiences, and in a matter of weeks he designed a turning lathe capable of making tips for ballpoint pens precisely and economically.

Putting the idea into practice did not go quite as smoothly. The Argentine technicians were willing to implement the design, but in the true spirit of "*mañana, mañana*," they did not even begin working on it. Meyne had another Spanish maxim up his sleeve:

"*Mañana es nunca!* Tomorrow never comes!"

Láng recommended a technician of German extraction for the job, hoping that he would be more reliable, but he had adapted all too well to the local way of doing things, and was many months late in completing the task. The other machinery could only be calibrated once the turning lathe was finished, and the task of enlisting suitable workmen was also delayed.

During the months of waiting, Bíró set up a chemistry lab in the kitchen, where he continued to try to perfect the material for the dye. He would regularly receive formulas from his brother, and try them out with little success. Even in the world of chemistry, the elder Bíró continued to display the industry of the young model student he had once been, but again proved incapable of developing his own ideas. Like a talentless dancer who learns only one movement and repeats it again and again, he would find it impossible to stray from the models outlined in specialist literature, mixing only various types of resins and oils. His younger brother put less and less faith in his recommendations, increasingly turning his attentions to glycol compounds—which would, in due course, become the ingredients of the ballpoint dye paste.

Mrs. Láng, who insisted on having her say in everything that went on around her, had a word or two of advice about the dye experiments. One night she made a call to Bíró, who was bent over his laboratory bowls.

"Put everything aside, Laci. I've found the solution to the perfect dye."

"And what's that?"

"You'll see soon enough—my chauffeur is on his way with it right now. All you have to do is color it blue," and with that she put down the receiver.

Her chauffeur soon delivered a scented white cream of unknown chemical constituents. As instructed, Bíró painted it blue and used it to fill a few pen cartridges. The following morning Mrs. Láng tried it out herself, but succeeded only in putting oily blots on the paper.

"Mari, now could I ask what exactly that was?"

"Before going to sleep I covered my face with my night cream, and it seemed so slippery I thought it could be transferred from pen to paper. We all make mistakes—it's only human."

The year 1940 had been one of pressured toil, and though the experiments had still not reaped tangible results, Bíró was confident that within a few weeks he would be able to produce working pens that could be put on sale. In early January 1941, however, Mr. Láng called him into his office, only to wave a pile of bills in his face.

"Mr. Bíró, you have gone over every imaginable budget. I can't finance this any longer."

"You want us to stop now, when we are on the brink of victory?"

"My name is Láng, not Rothschild, I'm afraid. Work things out as best you can, but don't count on me any longer."

Meyne nodded when he heard the bad news.

"You have to learn, old man, that a capitalist is a gentleman only up to a point—overstep that point, and out comes the thief in him."

But Meyne's latest words of wisdom fell on deaf ears; Bíró was deep in thought, trying to see the way out of this mess.

"We're not the ones affected by this," he said, "at least not immediately. We have enough money for bread and milk for a week or two."

"And beer," observed Meyne.

"And for beer. The biggest problem is that we can't pay the workers."

"Listen, Laci, I know you hate asking anyone for anything, but just this once, maybe you should force yourself. You are not these workers' boss; you are their friend. It's no accident that in the city this factory is referred to as the sanatorium. Get up there in front of them and try to talk this through."

Bíró accepted Meyne's advice, and the following morning he called the workers together.

"My dear friends! I don't know the Bible well, but I distinctly remember the following sentence: 'The laborer is worthy of his wage'. You, too, cannot live without your salaries. You have to give your families food to eat, and at the markets they don't give things away for free. I have an uncomfortable announcement to make. Right now I am unable to pay you. It's out of my hands—my partners want to pull out, as they have lost all their faith in the ballpoint pen, and want to stop flogging a dead horse. I don't share their views. I am absolutely certain that we can put the first batch on the market within a month, and then all our financial woes will be solved. I would like to ask you to work for deferred payment for this month. Of course, only those of you that are willing. If you aren't, I more than understand, and I cross my heart that I won't be angry. Now I will leave a piece of paper here for you. Whoever is willing to take the risk should put their name down on it."

And with that, Bíró swung around and went out into the corridor; for he did not want the workers' decision to be influenced by his presence. He felt it dishonorable to listen in on the conversation taking place in the workshop. He took a few more steps, tried to distract his attention, lit a cigarette, and mopped his brow with his hand. As he had always done in difficult moments, he tried to remember the shirt painted with dragon blood, but on this occasion he failed to derive the same energy from it as at other times.

"I've had luck on my side way too long," he thought to himself. "It was bound to wear off sometime."

The door to the workshop opened, and the foreman emerged holding the sheet of paper Bíró had left for volunteers to sign. The inventor took a quick glance at it: it was blank. He sighed, trying to face failure gracefully, and extended his hand to his employee.

"I quite understand where you are coming from. Maybe in your position I would say no, too."

The foreman grasped Bíró's hand in both of his.

"You misunderstand, Mr. Bíró. We have discussed the situation, and all forty-four of us have decided that there is no need for us to sign. We trust in you, and give you our word that we will keep working. If one month does not prove long enough, we are willing to make it two."

Bíró felt that he had just won the most important battle of his life, and silently apologized to his mascot, the shirt with the dragon blood stripe.

The inventor's luck continued to hold. It took only three weeks for the first, marketable model of the ballpoint pen to be completed.

The Lángs commissioned the Argentinean company Wolf and Partners to deal with the marketing and distribution of the pen. Wolf had a comprehensive network of commercial contacts and adequate experience. A wide-scale publicity campaign began in earnest.

It was the "eternal life" of the innovative product that gave it its name, Eterpen. The two-page newspaper advertisements depicted the ballpoint with two wings on its sides, as if to suggest that it could be used both on the ground and in the air. The latter characteristic was important, because Argentina, with its enormous territory of more than one million square miles, was witnessing a striking expansion in air travel. On airplanes traditional fountain pens were of no use for they would always leak as a result of the pressure difference. It was no accident that the airline companies were among the first purchasers of the new pen, making substantial orders and paying their bills on time. Bíró was insistent that the incoming funds first be used to pay the workers' outstanding paychecks.

The advertising slogan was "The Eterpen is two in one: pen and pencil"—though in Bíró's opinion this was taking things a little too far. He had no delusions about the quality of the first batch of pens. It was only its novelty that would make the product appeal to consumers. In reality, it was hardly past the experimental stage and a drop in sales could be expected at any moment. Yet Bíró's optimism was given a boost, and he no longer thought it unreasonable to conceive of the ballpoint replacing traditional writing instruments.

CHAPTER 18

During the summer of 1941 the newspaper *Hungarians of South America* published increasingly dire reports of developments at home. Following the provocative bombing of Kassa (today Košice, Slovakia), Hungary declared war on the Soviet Union, and the forces of the Hungarian army crossed the border. Forty-five thousand soldiers joined the fighting on the side of the axis powers. Domestic Hungarian politics was also increasingly pledging itself to the ideals of Nazi Germany; on August 8, Parliament passed the third and final Jewish law.

Bíró took every opportunity to remind the Lángs to honor their contractual obligation to arrange for his family to move to Argentina. They were highly elusive on this point, always devising distant and vague deadlines.

"All good things come to those who wait!" Láng comforted Bíró.

The success of the ballpoint pen began to have its effect on them, however, and on this occasion it was Mrs. Láng who approached Bíró on the matter.

"Laci, how much is getting your family out of Hungary worth to you?"

"But I've paid for it once already, when I agreed to sign the profit-share contract. The conditions were shameful."

"That was then, this is now, circumstances have changed. I'll ask you again: how much of your share of the profits are you willing to give up?"

"Is that a serious question? Whatever it takes."

"I think we'll have to specify the amount a little more precisely than that."

"Name your price, and I'll accept it blind."

Mrs. Láng shook her head, acting as if she were sorry she had brought up the subject.

"Maybe I have got too carried away. It would take years for you to pay back the costs. Travel, accommodation, immigration fees, for your wife, your daughter, your mother, your brother. No, maybe it is all just too much."

Her husband stopped her in her tracks.

"Mária, we are human, we have to do what we can to help others. I have faith in Laci's talent; I hope that in due course the ballpoint will conquer much more of the market. I would be happy to accept some of his share of the profits instead of money. Not too much, let's say half of his current share, 16.5 percent."

This offer meant that the inventor would own but a sixth of the total shares in return for his patent and three years of hard work. He accepted it without a moment's hesitation.

"I still think that it was the best deal I struck in my whole life," Bíró later commented. "In return for numbers on bits of paper, hoped and imagined future emoluments, I was given real human lives, the lives of those who meant more than anything to me."

Although he had succeeded in acquiring the necessary funds, the implementation of his plan continuously found itself up against obstacles. Although Argentina had not yet declared war on the Axis powers, to which Hungary belonged—this would only happen in the last days of the war, in March 1945—it nevertheless took a hard line on those wishing to emigrate from this region, and was reluctant to grant them visas. Bíró's claim was rejected, and he turned to General Justo as his last hope.

"My dear László," said the general, "you know only too well how much I would like to help you. But I am not above the law. There is no way they will let the family of a Hungarian into Argentina. But there is a solution, I think, if you catch my drift."

"I should be a dual citizen?"

"That is not a status recognized by Argentine law. You would have to become an Argentine citizen, and an Argentine citizen only. Would you be willing to do that?"

"Yes." Bíró was worried that his reply could have appeared forced, and so he added, "Of course!"

"Then get in touch the Hungarian ambassador here, Mr. Arnó Bobrik, and ask him to discharge you from Hungarian citizenship. He is obliged to give you a certificate of this. If he makes a fuss, tell him I sent you. Once you have done that, I will take care of the rest."

Bíró wasted no time in arranging an appointment at the Hungarian consulate, but it was weeks before Bobrik was willing to see him. Bíró was standing around in the street at the allocated time and he rang the bell, but it was a while before he was let in. As he was waiting, the door to the garage opened and a Mercedes pulled out. Bíró was able to recognize the Hungarian ambassador to Brazil: the younger Miklós Horthy, son of the regent and admiral by the same name who had presided over this "kingdom without a king" since 1919. Bobrik was sitting next to him. An overweight man brandishing a tennis racquet wandered out after the car and grudgingly gestured to the guest to come inside.

"Mr. Bíró? My name is Semsey, I'm the ambassadorial counselor. The ambassador has asked me to deal with your request. I am afraid I have some urgent business to attend to as well, so I hope we can get this over with as soon as possible." He pointed to a bench in the embassy garden. "Maybe we could take a seat out here."

Bíró had difficulty swallowing this condescending manner, but did not allow his feelings to get the better of him. He asked for the certificate of absolvement. Semsey frowned.

"Are you so easily influenced by the purported advantages this would offer?"

"These are not purported advantages; this is escape for my family."

"This is your homeland you want to wave goodbye to!"

Bíró pursed his lips.

"In my experience, no one ever makes reference to their homeland, nation, or people if they want to give something. These concepts are always used to ask for something or demand something. I am unmoved by this line of argument, I'm afraid. I am trying to forget Europe, with all its blessings and all its curses."

"And you think that Argentina will change things for you?"

"No, my home will not be Argentina. My home will be 3031 Calle Buschiazzo, where I live, and where, in due course, I hope my family will live. That's all the change I need."

Semsey looked at his watch, and stood up from the bench.

"I have taken note of your request, Mr. Bíró. I will pass it on to the ambassador, and we will inform you of our decision presently."

"I would like an immediate answer. As I say, my family's fate is in the balance."

The counselor looked up.

"How dare you use that tone of voice with me?"

Bíró took out Justo's business card.

"The general told me that it would possible to deal with the matter urgently."

Semsey inspected the card, read the recommendation Justo had written on the back, then gestured reluctantly for Bíró to follow him into his office. Within a few minutes Bíró had the certificate he needed in his hands.

He set off for home on foot, angry with himself that in the heat of their discussion he had fallen for Semsey's provocation instead of saying what he had wanted to say. In retrospect, Bíró was embarrassed by his strong words, and that he had not spoken of the thousands of pleasant memories he would always associate with Hungary. Some of those remembrances flashed up before his eyes: the playground of his youth, his father's smiling face surrounded by the crowd at the coffee-house; the rain-swept nighttime walk around the Vajdahunyad castle in Budapest's City Park when he had asked for Elza's hand. He still had the familiar sweet taste of chestnut purée in his mouth.

"Fate may still punish me for this," he thought to himself, shook his head, and quickened his steps.

Informing his family of the news was no easy task. As the scale of fighting grew, Hungary became excluded from the American-controlled airmail service, meaning that contact was lost from one day to the next. It took a while for Bíró to find out that the International Red Cross would accept deliveries to Hungary, but only postcards of twenty-five words—for ten times the normal rate.

Of all his friends and acquaintances back in Hungary, it was Imre Gellért that Bíró trusted most. He sent a series of these short cards to his old friend, asking him to help obtain passports and visas for the members of his family.

There were times when this line of communication was broken, and Bíró was unsure whether they had succeeded in escaping from war-torn Europe, or whether his rescue mission had been too late. A great weight was taken off his mind when he finally received the news that all four of them had safely left Hungary. There was a fifth member of the party: a young boy called Róbert Vermes, whose divorced mother sent him to his father in Argentina. The train journey to Bilbao took three days, throughout which they were in a locked compartment, unlit because of the wartime blackout. The elderly Mrs. Bíró slept on the seats on one side, the two children on the other, with Elza and the elder Bíró forced to lie on the floor.

On arrival in Bilbao in September 1941, they collected their boat tickets, and after a wait of four days, boarded the Spanish ship *Cabo de Hornos*. Though the steamer was built to house a hundred and fifty people at most, there were more than three hundred crammed on board. The Bíró family had to squeeze into a cabin built for two.

The *Cabo de Hornos* succeeded in leaving the coast of Europe, but it had by no means yet reached safety. The obsolete steamer needed three weeks to cross the Atlantic, and as it made its journey across the open waters it was continuously exposed to serious threat. Both the Allies and the Axis Powers were known to attack ships sailing with neutral flags.

When they arrived in Barbados, the British authorities made nine of the passengers disembark—including György Bíró—on the suspicion that they were working as spies for the Germans. After they were questioned, the prisoners were put in a local internment camp, and the Bíró family had to continue its journey without him.

The threat of attack was still in the air as they approached South America: only a few weeks previously, the Hungarian boat *Kelet* had been torpedoed just off the Argentine coast. The whole odyssey lasted twice as long as originally planned; the boat arrived in Buenos Aires forty days after its departure. Now, for the first time in a year and a half, Bíró was able to hold his family in his arms again.

It was another two weeks before György Bíró joined them. The British had found the spy among the prisoners on Barbados, and he was taken aboard a military ship and executed. They apologized to the others and allowed them to continue on their way. György, his face tanned, disembarked in a good mood.

"*Buenos dias! Què tal?*"

The inventor clapped his hands together.

"You already know some Spanish?"

"Don't be silly, I've just learned the most important phrases."

"Will you ever grow out of being teacher's pet?"

CHAPTER 19

Once settled in his new surroundings, György tried everything he could to step out of his brother's shadow and make
a living of his own. He inquired as to what use he could
make of his Hungarian dentistry qualifications, only to be told that to
practice he would have to start his studies again from scratch. There
were regular reports in the newspaper for the Hungarians of Buenos Aires about qualified engineers or doctors successfully passing
elementary school exams. Resigning himself to his fate, György gave
up his ambitions to be independent from László, and started work on
mixing the components of the dye in the factory laboratory. He did
not insist on pursuing his earlier ideas for the formula, and followed
his brother's intuitions, with considerable success, produced increasingly impressive samples.

Freed at last from the burden of the chemical experiments, László
Bíró could turn his attention to the mechanical development of the
pen itself. He had his work cut out for him, as customers returned
more than half of the pens with a list of quality complaints. The ink
would sometimes flow backward, behind the plunger, or soon harden,
blocking up the cartridge.

The company did its utmost to maintain the reputation of its
product. At Meyne's recommendation, a beautifully decorated service
center was opened in central Buenos Aires. Pretty hostess girls were
on hand behind the desks to exchange faulty pens. They also introduced a guarantee that provided free dry cleaning vouchers if the pen
stained a customer's jacket or trousers. These measures were not only

expensive but also generated a sort of bad publicity in their implicit suggestion that the product still required further development. A fundamental solution to the problems was what was required.

Bíró was at a loss to understand why, when they had been produced with the same technology, some of the pens worked perfectly, while others were defective. He explored every minute detail of the design to look for the possible source of the fault. First, he examined the writing ball. It was made of steel, about one millimeter in diameter, and its surface had been artificially roughened so it would cling to the paper better while writing. This component was supplied by the legendary Swedish company SKF, which would seem to guarantee flawless workmanship. Bíró was never a believer in legends, however. He ordered a measuring device precise to the nearest thousandth of a millimeter, and used it to grade the precision of a few dozen balls. His doubts were vindicated by the results: though many of the balls were perfectly spherical, a large number were slightly elliptical.

Bíró tried out the differently shaped balls in separate pen bodies to compare how they wrote. In the case of the perfectly spherical ones, the pens almost always wrote perfectly while the others leaked or drew broken lines. Having thus found one of the sources of imperfection, Bíró immediately registered a complaint to the SKF Company. The Swedes promised to send specialists out to the factory, but their journey took many long weeks, and as they could not stop production for such a long time, Bíró had to find a provisional solution to the problem.

Tens of thousands of balls had been collected in the warehouse and it would have been impossible to check each one individually with the measuring device. Bíró could not hope for any further investments from the Lángs; they already regarded the project as having gone over-budget. Forced yet again into a corner, Bíró had to rely on his invention, devising a solution that was as simple as it was ingenious: "I had a large sheet of metal cut and polished until it gleamed like a mirror. I placed the sheet on my desk at a slight angle of fifteen degrees. I attached a feeding apparatus to the middle of the top of the sheet, which sent the balls down the slope one by one. Under the bottom of the sheet I placed a number of different boxes: the perfectly round balls would always fall into the middle one, the slightly distorted balls into the outer ones." The success of this selection process was born out by testing of the pens, but it was too labor-intensive to be efficient in the long run. Fortunately the Swedish engineers arrived in the meantime, and after thorough tests accepted Bíró's complaints.

The excuse they gave for the fault was that no previous customer had ever needed such small but perfectly formed balls. To preserve the company's good name, however, they promised to start a new production line on their return to Sweden and make changes to the technology employed. Within a few months the quality of the balls had indeed improved: ninety percent of the balls in the next shipment proved to be perfectly spherical.

Bíró, a master of practical solutions, also used another device to improve the efficiency of this quality control process. He adapted a discarded record player, relocating the platter off-center. He attached an arm that moved a ballpoint pen around in a circle; underneath the pen a piece of paper moved round at a controllable speed. As the pen drew a line on the paper, it was possible to judge whether it was continuous or broken. It was this apparatus that accidentally brought about a great leap forward in the development of the perfect ballpoint pen.

One evening, in the fifteenth hour of his working day, an exhausted Bíró was going through the motions of his experiments almost robotically. His attention would often wander. While testing the machine he regularly had to stop it to push the ink in the cartridge toward the tip of the pen with the plunger. The last time he did this, his mind was again on other things for a moment. Forgetting the plunger, he put only the tip and the attached cartridge on the testing machine, then locked up the laboratory for the night. He realized his error only when he pondered the results of the day before going to sleep, but did not trouble himself with it too much.

The following morning Bíró started to change the pen held in the turntable test machine, only to find that the tip and cartridge had been capable of writing continuous circles without need for the plunger. To his surprise, the last drop of ink in the cartridge had been used up.

Bíró's world turned upside down. He was now in the fourth year of his experiments on the ballpoint. During this time, he had concentrated his energies largely on the ball itself and the composition of the dye. He had taken the use of the plunger for granted; it had not crossed his mind that it might be possible to leave it out of the design altogether. It was serendipity, the element of chance so often needed for an invention to come to fruition, that proved to Bíró that the pen could work without a plunger and with its ink cartridge open.

But this lucky discovery did not put Bíró's mind at rest—quite the opposite. He continued to get a mere three or four hours of sleep each night. In practice, he now lived in his workshop, smoking two packs of Craven 'A' brand cigarettes a day.

Bíró attempted to compare the phenomenon to the laws of capillary action, filling tubes of different sizes with liquids of varying viscosity. If the right liquid is poured into a tube of optimal diameter, and one of the ends closed, a siphon similar to the kind of pipette used in the chemical industry is created. The special characteristic of this tube is that atmospheric pressure does not allow the liquid to flow out of its open end, regardless of whether it is held in a horizontal or vertical position.

Bíró attempted to apply this principle to the construction of the ballpoint pen as a whole. The task was not without its difficulties. What was certain was that the writing end of the ink cartridge would have to be sealed with a vacuum for the capillary process to take place. Such a solution would present technical difficulties, however, as there had to be a slight gap between the ball and the socket for it to be able to rotate. It would take many more months of experimental work for Bíró to discover that if the dye had the necessary viscosity and surface tension, the movement of the ball would not be impaired.

CHAPTER 20

The ballpoint pen that was the final result of these continual modifications bore little resemblance to the first model Bíró had described in his patent registration application submitted to the Hungarian Royal Patent Court on April 25, 1938. He was now witnessing the birth of a pen that would satisfy the customer's every need, and that required a patent application in every country where it might have a significant market. The huge costs of such a process were out of the question for the Lángs, however, as they would have been forced to bring in another partner. Meyne offered himself for the task.

The former Hussar officer had already established a blooming network of contacts in Buenos Aires. For example, he had found a place in the city's most prestigious equestrian club, the El Caballero, where he taught amateur fans of the sport. One of his students was Leon Hartung, director of the Argentinean office of Philips, whom he wanted to win over as a business partner.

Meyne had picked up enough technical and commercial expertise to make a convincing case for the opportunities offered by the production and distribution of the ballpoint pen. Hartung was willing to get involved in return for exclusive rights for Philips to distribute the pen throughout South America. From time to time he would transfer sums of money, which Bíró would call their "Hartung injections." Having a contract with an international company also revived the confidence of Mr. and Mrs. Láng, who were in turn willing to make further investments.

Bíró pressed on with his experiments. New ideas were to be born of the long hours of work: the writing tip, previously fixed in one place, could now be made to pop out of the body of the pen with the push of a finger, then made to pop back after use with another click. Making good progress, Bíró was certain of his imminent success, but after a few months the financial sources he had thought to be so dependable wavered.

After the invasion of the Netherlands, the headquarters of the Philips & Co., in Eindhoven, fell into German hands, and the Nazis restructured production to serve military needs. Their control came to extend to the foreign divisions of the company, including that in Argentina. They did not wish to work on the ballpoint pen, a product that had not even been introduced, and sought an excuse to back out of the stipulations of the contract. They produced a manipulated market research survey showing that, at most, sixty thousand units of the pen could be sold. And this figure was based on the assumption that the price of the pen would not be higher than five dollars and a half—while the Lángs were planning to sell it for twenty-five dollars. Hartung, meanwhile, who had signed the contract and who continued to support the project, was transferred to Canada—and his successor made no bones about voiding the agreement.

The company was unable to compensate for the loss of income incurred by Philips' withdrawal. The work at the factory was running up higher and higher bills, and the Lángs soon used up their last reserves. They took out a loan from a bank, stating that it was for the production, not development, of the pen, in order to enjoy a lower interest rate. Unfortunately, when the time came for repayments to begin, they were unable to present a perfectly working model of the product, and the bank was unwilling to extend the deadline.

For the second time during the course of the experiments, the Lángs and their partner Müller announced that they would be unable to pay weekend overtime. Bíró again offered to intervene personally.

"I'll try talking to the workers, and perhaps they will listen to me again."

"On this occasion you would be getting their hopes up in vain,

Mr. Bíró. Even were you to succeed, it would make no difference," said Müller. "We cannot afford to buy the materials needed for production to continue. We haven't paid the outstanding bills yet and none of our suppliers will give us credit."

"We can't just stop now!" Bíró objected. "I give you my word that in a matter of months the ballpoint pen will be an international sensation. We must grin and bear it until then."

Láng pursed his lips.

"You know where we'll be in a few months' time? We won't even have beds to sleep on. Let's be honest about this: I have every respect in the world for your talent and your hard work, Laci, but you are a born loser. If your job was to sell candles, the sun would never set."

Mrs. Láng added vehemently:

"You should never have gotten us involved in such a dubious undertaking. You have ruined two people."

"Three people!" said Müller.

Meyne, who had been observing the discussion in silence, raised his hand.

"Excuse me for interrupting, but I think there is an answer to this. I am reminded of a tactical rule from my days in the officers' corps: you must always make your escape from the smoke of battle forward, not backward."

"Don't talk in puzzles, man, what are you saying?"

"It would be absolute suicide to call a halt now. To the contrary, we must invest all our energies in increasing the pace of our work. We have to start a propaganda campaign for the pen the likes of which have never been seen in Argentina. Our posters must be on every street corner."

"You still rest your hopes on the consumer? We know from experience that their interest is no substitute for capital."

"It's not them I'm thinking of," explained Meyne, "but the 'big sharks.' You know as well as any of us that capital has fled out of war-torn Europe. And it is in the very nature of capital to find work somewhere. It holds out a torch, looking for the places where it can make itself useful. I have information to suggest that the Argentine economy is overheating, what with supplying grain, meat, and mineral products to both sides in the war; it has the highest foreign currency and gold reserves in the world. The National Bank has accumulated so many bars of gold that that's what the guards sit on instead of chairs. All we have to do is to draw investors' attention to ourselves."

Mr. Láng stopped Meyne's diatribe in its tracks.

"That all sounds very good, but how do we pay for this public relations campaign?"

"For the time being we don't pay for it. The advertising expenses only have to be settled six months after the ads appear. The newspapers will welcome us with open arms, for no editors can imagine a better article than a three-column advertisement. Six months is a long time. Laci says that we will see the necessary results before then; you have said that you have faith in his talent. I can't see what the problem is."

"And if it doesn't work out, after all?"

Meyne shrugged his shoulders.

"Then we have come a cropper anyway, and it really doesn't matter whether we are strung up in five places or in six."

Láng objected:

"This whole thing is a great big bluff. A castle in the sky for which we don't even have the key."

"Maybe, but no one can tell the difference between it and any other castle."

The practically minded Müller had further worries.

"How do you envisage that production will continue? As I said, we don't even have enough money to make a visit to the supplier of the raw materials."

"Who wants to visit them? They can come here. They will bend over backwards, grant us unlimited credit, just as long as they don't lose a world-famous client."

Everyone looked to Láng for his response. The businessman nodded.

"Beggars can't be choosers. We have no option but to accept Meyne's suggestion, however touch-and-go it might seem."

After the meeting drew to a close, Bíró and Meyne walked home together, as was their custom.

"There's one thing we forget to get straight," said Meyne. "What are we going to call the new pen?"

"Can't it just be Eterpen again?"

"No, we've flopped once with that name already and the bad associations won't go away. New brooms sweep clean, as they say."

"Do you have something in mind?"

"I would have one suggestion. Your name has such a good Latin ring to it, and it's pretty short, too: why don't we call it Biro? It rhymes with "bureau"—desk—which is where it will be put. 'Bureau' is also what the French call an office, and it's something similar in Russian as well."

"Thanks for thinking of me, but that's not just. I would never have got this far without you—I would still be in a German prison in Paris. Let's call it Biromeyne."

"That's too long! If you deserve two syllables, I only deserve one at the most!"

"Birome, then?"

Bíró repeated the name again and again.

"Birome, Birome, Birome! I like it!"

CHAPTER 21

The following day Meyne sprung into action. He visited El Globo, the best-known advertising and public relations firm in Buenos Aires. He planned his arrival carefully in advance: Müller drove him to the office in the Lángs' big Packard, and when they arrived, Meyne asked him to get out and open the door for him, to make it clear that an important individual had arrived.

Meyne was welcomed by the owner of the company, Donna Maria Fulgencia, a widow of Italian origin who was entranced by the former Hussar officer's manners, and readily accepted the task of arranging the promotional campaign.

Everything went according to Meyne's original plan. Double-page spreads in the newspapers heralded the "radically new ballpoint pen that works perfectly every time," and the radio and film news were quick to make a visit to 3040 Calle Oro. To prove its resilience, the body of the pen was thrown down deep valleys and run over by buses. Rented airplanes wrote the name "Birome" high in the sky above Buenos Aires. Meyne also leaked a fake report that the peace treaty that would end World War II would be signed with a Birome pen. Bíró, however, was a trifle unnerved by these publicity stunts.

"Do we have to make such a big to-do over this?"

"If we want to win, we do."

"And all this doesn't bother you?"

"No. It we do have to fail, we might as well fail in the view of the whole world. There's nothing worse than being executed in secret."

László Bíró as a young man

The automatic gearbox

Bíró's press card and journalist's accident
insurance ID

Bíró with his motorcycle, before
departing on a trip to Germany

Swiss patent certificate for the ballpoint pen

Hungarian patent certificate for the exchangeable-cartridge ballpoint pen

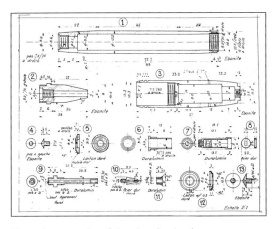

From the appendix of sketches for the Argentine patent for the ballpoint pen

Patent description for the electro-magnetic mail dispatch system

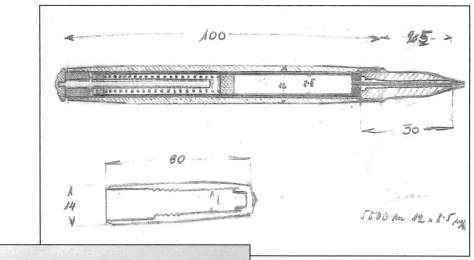

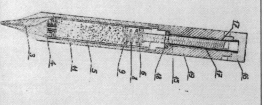

Bíró's technical drawings of the ballpoint pen

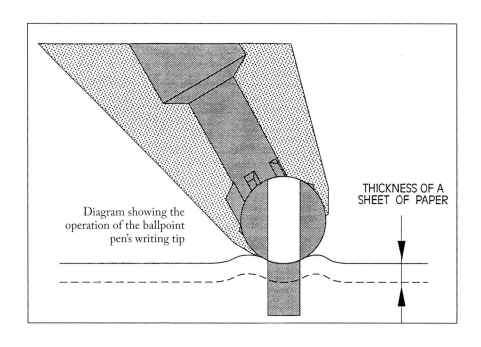

Diagram showing the operation of the ballpoint pen's writing tip

THICKNESS OF A SHEET OF PAPER

Nyugati train station, Budapest, 1935

Bíró's boat ticket from Barcelona to Buenos Aires

Buenos Aires from the period of Bíró's arrival

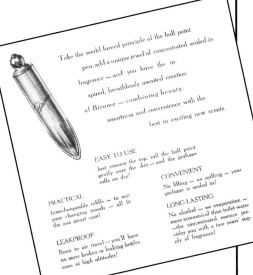

Take the world-famed principle of the ball point pen, add a unique jewel of concentrated sealed-in fragrance — and you have the inspired, breathlessly awaited creation of Biromé — combining beauty, smartness and convenience with the best in exciting new scents.

EASY TO USE
Just remove the top, roll the ball point gently over the skin — and the perfume rolls on dry!

CONVENIENT
No filling — no spilling — your perfume is sealed in!

PRACTICAL
Interchangeable refills — to suit your changing moods — all fit the one smart case!

LONG-LASTING
No alcohol — no evaporation — more economical than toilet water —the concentrated essence provides you with a two years' supply of fragrance!

LEAKPROOF
Boon to air travel — you'll have no more broken or leaking bottles even at high altitudes!

Newspaper advertisements for the Birome pen and for ball-point perfume

conoce la satisfacción de un
GRAN REGALO

con la
ESFEROGRAFICA*

INDUSTRIA ARGENTINA · RETRACTIL · *MARCA REGISTRADA

EXCELSIOR

en cualquiera de sus espléndidos modelos
PARA ESCRIBIR MAS Y MEJOR!

BIROME S A. Alsina 633 - T. E 33 - 5075 · Bs Aires

29

Revolves on a Ball-Point

Big Men of the Pen World

Magic names in writing are these, l. to r., L. E. Waterman, who made the first practical fountain pen; Ralph K. Bard, chairman of Eversharp; Milton Reynolds, chairman of Reynolds International, and Laszlo Biro, who developed modern ball-point pen. At right a couple of early pens, on self-explanatory cards. Pen upright at left of picture is an early Waterman. In box is $125 modern pen-pencil set.

Pen by H. K. McClelland
Patented April 17, 1855
Pressing the key · indicated by arrow · allowed ink to saturate the sponge which fed it to the nib.

Patent No. 648
March 21, 1838
Inventors · H. C. Windle
J. Gillote · S. Morris

fountain pen, would work as smoothly as the rotary press. The ball-point was not his own idea. That had been tried as long ago as 1880, but nobody could master the precision machining necessary for a workable pen. However, Biro took that idea and pushed it around.

Laszlo and Georg made their first model of the pen in 1936. In 1938, a French firm, Elveco, began making pens from this model. But the war was creeping up on Biro, and he fled Hungary to France, and finally, France to Argentina. There, with a capital of 30,000 pesos, he founded the Eterpen company manufacturing the Birome (pronounced Bsero-may)-ter, Maria, now 14, joined him in Argentina and he settled down as a pen manufacturer.

Army Sold on New Style Pen

The Birome pen became well known in South America which is very fountain-pen conscious. In

quire retooling and a new approach toward writing. However, the Army made up their minds for them.

Biro's representatives had previously approached most of these firms, who hadn't been receptive at the time. But now the firms approached Biro. Eberhard Faber, for generations a pencil outfit, and Eversharp made a deal with Biro for American rights and commenced developing a North American product based on his design.

They worked quietly, under conditions of top secrecy, something like the Manhattan Project. But nothing much goes on in the fountain pen business without everyone hearing about it. If there is a new development in the Parker factory in Janesville, Wis., it is soon known in the Sheaffer plant at Fort Madison, Iowa, at the Faber plant in Brooklyn, at Waterman's in Manhattan and at Eversharp's in Chicago, to say nothing of the little hole-and-corner loft manufactories scattered all over. It's a good thing the internal espionage

from Albert Lea, Minn., where his father had a farm. The family moved to Illinois and Milton renounced formal education at the age of 13. Like others in his line of work, Reynolds is bulging with energy, which squirts out in all directions. Some of this he used to work off by long-distance swimming. Flying has now taken place of swimming.

He ran rapidly through a series of business enterprises, many of which ended in bankruptcy, but was never daunted. He took pride in his vision and in his lightning decisions, right or wrong. It didn't matter much to him what he was doing as long as he had something to sell, preferably something that could be moved rapidly and in large quantities.

He found a steady backlog in a machine that prints showcards and pricecards, but was always ready to prove his prescience with a new venture. One of these was Mexican silver cigaret lighters. At a time when lighters were short in the U. S., Reynolds contracted for the output of a lot of household factories in Mexico, shipped them

sharp Co. The latter made automatic pencils and fountain pens for years, but got itself enmeshed in complicated engineering dreams and was in financial straits in 1939, when Martin Straus 2d and Ralph K. Bard took over.

Straus and Bard, who was Assistant Secretary of the Navy for quite a while, had been associated for years with investment firms. When they became interested in the Wahl project, they decided not only to invest but to participate actively in the reorganization of the concern. They brought in a lot of veteran fountain-pen men, and notably Larry Robbins, sales manager and senior vice president, who had been in pens for most of his life.

Eversharp, like Sheaffer, Parker, Waterman, had large investments to protect and considered it ungentlemanly of Reynolds to jeopardize these investments with a project begun on a $26,000 shoestring. Choking down their anger, however, they proceeded as planned. Eversharp brought out its ball-point—excuse it, ball-point is an phrase used at the Eversharp of-

for any one of several reasons. Another is that, it has been reported in various quarters, the ink will fade, thus leaving a document unsigned or unreadable. These two items alone are driving the manufacturers crazy.

Along with this piece is a schematic drawing of a ball-point pen, in this case the Eversharp, which the company prefers to call the CA pen, CA standing for capillary attraction, the principle on which the instrument functions. The thing looks simple and is, and it seems impossible that a simple instrument could require so much study and experiment.

In the Eversharp, as in the Birome, the Eberhard Faber, the Biro principle, the ball, a millimeter (about one twenty-fifth of an inch) in diameter, is seated so that it revolves freely in all directions but with an extremely fine tolerance. This requires precise machining, which is no particular problem these days. Back of the ball there is a duct which carries the ink from the cartridge to the

Newspaper article on the ballpoint pen (Bíró on the far right among those pictured)

Argentine stamps depicting the country's great inventors, with Bíró in the top left with the ballpoint pen

Andor Goy as a young man

Goy with his colleagues

The "Gopen"

The Goy and Kovalszky typewriter and calculator shop

Andor Goy, later in life, in the factory of ICO Inc., one of Central Europe's largest manufacturers of stationery and writing instruments

The first result of the image-boosting exercises was evidenced by the company's suppliers, who needed no second bidding to send the necessary materials, and production geared up again. But the inflow of new capital, the primary objective of the campaign, was nowhere to be seen. Meyne increasingly made a habit of paying evening visits to Donna Maria Fulgencia to lick his wounds. He took her flowers and, on one occasion, a pen with her name engraved in it.

More than a week had passed since they had begun work on the publicity drive when one morning the telephone rang. The secretary of one of the largest financial institutions in Argentina, the London Bank, announced that Henry George Martin, president of the board, was inviting Messrs. Bíró and Meyne, together with Mr. and Mrs. Láng, to visit him at the offices of the bank at ten o'clock the following morning.

"The wall has come crumbling down!" said Meyne.

Martin received his guests in his office.

"We have made initial investigations into the ownership structure of your company," he began. "Is is true that a hundred percent of the shares are in your hands?"

"Yes they are, though we have received a number of offers for options on them," replied Meyne, who had nominated himself as the spokesman for the shareholders. "And we will choose the most agreeable one in the next couple of days."

"I very much hope that our offer will be able to compete with the best of them. We would like a year-long option on 51 percent of the shares, for which we are willing to pay 65 thousand dollars—that is, 325 thousand pesos—in cash. Naturally, we would negotiate the price of the shares at a later date."

"That is a favorable offer, certainly, but we are bound by honor to give our other partners a chance to make theirs. We would like forty-eight hours to think it over."

"I am forced to accept these terms, but of course they also release me from any obligation." Martin rose from his chair. "I will be expecting you the day after tomorrow at the same time. If by chance we reconsider our offer, we will inform you without fail, so you don't make the trip in vain."

They had hardly got in the car when Mrs. Láng let Meyne have it.

"You pathetic little man, you've ruined everything! We should have accepted the offer on the spot. Before we know it, the London Bank will back out and we can start digging our own graves."

Bíró knew his old friend well enough to know that he wouldn't take such an insult lying down. He feared a conflict, but Meyne kept the tone of his voice in check.

"Mária, you are a first-rate businesswoman. It cannot have escaped your notice that the sum in Mr. Martin's offer is not a round number."

"And so what?"

"My humble amateur opinion is that this is a signal that if we start to haggle, he would be willing to raise the offer somewhat. We should make use of that—but we have to keep him waiting. If you don't agree, you have every right to turn back and tell him that you accept the arrangement. You would make Mr. Martin's day."

Meyne's strategy was successful. The banker did indeed raise his starting offer, and in the end the Lángs received 75 thousand dollars, or 375 thousand pesos. This amount was beyond their wildest expectations, and more than enough to cover all the expenses the development of the pen had incurred. Moreover, it was essentially an advance: were Martin to exercise his option, they would receive payment for his 51 percent of the shares, while retaining the other 49 percent for themselves.

On September 29, 1943, Bíró woke early in the morning, as usual. He began the day with a coffee and a cigarette, then made the journey to work on foot. It seemed as if the porter's welcoming bow was a little more enthusiastic than usual, but he did not respond. His mind was on other things.

As he walked down the corridor he was met by the Lángs, Müller, and Meyne. They were all dressed up for some kind of occasion, and gave him a hug in turn, congratulating him. It was only then that the penny dropped in the inventor's head that it was his birthday. They guided him into the factory workshop, where the workers greeted him with wild applause.

"Hola, Ladislao!"

From one corner his daughter Mariann run up to him, carrying flowers, and after a few moments she was joined by her mother. Meyne pointed to what was laid out on the table.

"There you are, Laci! A little surprise for your birthday!"

He saw two huge number fours made out of marzipan, together with a cornucopia of sweets and fruit. In the middle was a silver plate, empty apart from a Birome and a document underneath it. Bíró lifted it up to have a look: it was the Argentinean Academic Council's certificate of the pen's excellent quality of manufacture. Bíró picked up the pen and sat down. For a moment, it was as if he were alone in the packed room. It crossed his mind that it was getting close to six years since he had first had the ballpoint idea. During that time every aspect of the world around him had been transformed: he lived in a distant country, under a foreign nighttime sky, fumbling for words in a strange language. He could not help feeling that he had paid too high a price for this victory. He would happily have snapped the pen in two, but he felt something pressing on his arm: his little daughter clung to him.

"Let me have it, Daddy!"

Bíró gave it to her with a sigh. He forced a smile, and took the glass of champagne Meyne was handing to him. Arc lights were suddenly switched on in the background.

"What's that?"

"The film news cameras are here. You'd better say a toast!"

"What should I say?"

"I'll leave that up to you."

Bíró struggled to gather his thoughts in Spanish.

"Now, as I think of all the things that have happened to me over the last forty-four years, I can only come to the conclusion that there is a defining order in Nature, an order that governs us all. We do what we do without knowing why we do it, and that is how our results come about." Bíró mumbled to Meyne from the corner of his mouth: "Is that enough?"

"No, go on!"

"My daughter once asked me: Daddy, who is more important, the porter or the president? I told her that they were equally important, because neither could exist without the other. There is a mutual, interdependent relationship lying behind everything, whether we happen to be aware of it or not. But she continued to quiz me: which of them is a good man? I replied: the one who is at peace with himself and with his surroundings."

As Bíró became silent, the reporter took his turn to speak.

"We have been told that you have more than forty inventions to your name. Which is the most important for you?"

"Always the next one. I would not like only to be associated with the ballpoint pen."

"Mr. Biro, please tell us a few words about the inventor's profession!"

"Actually I can't really tell you why I keep inventing new things, but if I look back on my life and ask myself whether I did it just out of a whim, the answer is no. I had to invent these things—that is how it had been determined. Fate wanted to use me as a medium to express itself. The most important thing is that we should not be afraid of what we don't know."

The reporter hazarded a final question:

"Could I ask what advice you would give to the young, to the inventors of tomorrow?"

"My advice is that they do not let life pass them by. They should always remain aware that life is the greatest pleasure we have."

Thanks to its omnipresent publicity, the Birome had become a commercial hit. Customers were willing to pay as much as eighteen dollars for one of the gold-plated, push-button pens. Much attention was paid to the fact that the United States Air Force ordered twenty thousand units to give to its pilots; whatever alterations were made to them, fountain pens always leaked at high altitudes, while the ballpoint pen worked without a hitch.

The success encouraged Henry George Martin to come to a rapid decision. He did not wait for his year-long option to expire, but used his right to purchase much sooner, paying the Lángs the price of one million pesos that they had previously agreed upon, and thus laying claim to the majority of the shares.

Mr. and Mrs. Láng called together the board to divide the million pesos in proportion to the division of ownership. Bíró was shocked to find that they reduced his share by the total value of the monthly advances he had been receiving to cover his living expenses, even though this cost had already been recovered from the money paid for the share option. Mrs. Láng did not leave it at that, moreover, insisting that her husband should receive some of the forty-nine percent of the shares that remained in the company's hands. This, Bíró was unable to take lying down.

"How can you justify that, Mária? Lajos has already received a three hundred percent return on the money he invested!"

"If my husband had not shouldered the financial responsibility, this invention would never have seen the light of day. Not to mention the fact that without his help neither you nor your family would have been able to escape from Europe."

Bíró had heard these arguments one time too many, but was still susceptible to the emotional blackmail and struggled to find counter-arguments.

"I have always been eternally grateful to Lajos for that, but I think I've repaid my debt many times over by now. Add up how much money I've been left out of."

"You think money brings happiness?"

"From the way you talk, it obviously does, Mária."

Bíró discussed the matter with his closest partners, his brother, and Meyne. They came to the decision that it was not worth arguing the point any further with Mrs. Láng. He agreed to having half of his portion of the shares—which, of course, was now only sixteen and a half percent—transferred to Lajos Láng.

"The 'good cop' deserves that much," said Meyne.

As a majority shareholder, Martin now took the company's affairs into his own hands with zeal. Because of the more favorable tax conditions in Uruguay, he set up a holding company called Fomento in the country's capital, Montevideo. He commissioned a legal practice in New York to certify patents for the pen, and held talks with a number of large companies in person.

His efforts were soon to bear fruit. One day there was a call to the office in Calle Oro from New York. It was the director of the world-famous Eversharp fountain-pen company, who wished to hold talks in his New York offices about purchasing the rights to produce the pen. The Argentine company was unanimous in mandating Martin as its representative; in addition to his superior financial expertise, he also enjoyed the significant advantage that, as a British citizen, he did not require a visa to enter the United States. Opinions were divided, however, as to how much they should ask for the production rights.

Their expectations were of a sum in the region of half a million dollars; some thought this figure too high, others too low, and in the end they left the decision up to Martin.

The banker traveled to New York, where he began negotiations with the vice-president of the Eversharp Company. A few days later Martin's partners back in Argentina received the news that he had achieved success beyond their wildest expectations, agreeing to a price of two million dollars for exclusive rights in the United States, Canada, and Puerto Rico. Bíró was, by now, wise enough to know that he would only ever receive a fraction of a percent of this profit.

"I often reflect that if I had only had a slightly better sense for things financially, I would have achieved unimaginable riches with my invention, but I feel no kind of resentment. It took so much struggle over so many years for me to make a success out of my idea. The ballpoint pen easily became the world's most popular form of writing instrument, overtaking the pencil and the fountain-pen, and this thought fully compensates for all the wealth I missed out on."

CHAPTER 22

After the performance was over, Gellért sat in the artists' lounge waiting for his wife to come out of her dressing room. She soon appeared, greeting her husband with a kiss.

"How did I do?"

"There wasn't a dry eye in the house."

"Flattery will get you everywhere, darling."

"You are my goddess! I love you!"

They set off into the night on foot. Their flat was only a few minutes from the theater, but on the way home they stopped for a night cap at the Luxor Café, which had just been restored and reopened after suffering damage in the war. They ran into some friends and chatted long into the evening; it was around midnight when they returned home. They had hardly taken off their coats when there was a ring at the door.

"Who's there?" Gellért asked.

"Captain McGray, a courier for the Allied Supervisory Committee. I have a package for you."

The memories of the nighttime visits of the Gestapo were still fresh on Gellért's mind, and he only opened the little window in the door. He saw the American soldier in the corridor outside give a salute.

"Mr. Gellért?"

"Yes?"

"This is for you."

He handed him a thick packet, asked Gellért to sign his book, and left.

"Don't open it, it might be a letter bomb!" cried his wife.

"Don't panic, Olly, this isn't a French melodrama!"

Gellért looked at the name and address of the sender:

```
Compania Sudamericana Bíró.
Telegramma: Birocco, Telefono: 73 Parque 4709
Buenos Aires, Calle Oro 3040
```

"It's from Laci!" he said excitedly, opening the parcel. The first thing he took out was a business card in an envelope, in which was hidden a hundred dollar bill—which amounted to a small fortune in Budapest after the war.

"Not a bad business card. Olly, hide it somewhere safe, we can't have anyone finding it here."

He then took out an elaborate little box. Inside it was a ballpoint pen; below the Birome marking were engraved the words: *To Imre, from Laci.* For a moment Gellért saw stars; he was completely subsumed by his delight.

"My God, he did it after all. It's the ballpoint pen! Let's toast this with something special, Olly!"

"What did you have in mind?"

"There's still some cognac left from your première, isn't there?"

As she went to get the bottle, Gellért tried drawing various figures with the pen on a piece of paper. It worked perfectly, with no gaps; neither did it leave so much as a trace of ink on his hand. Gellért lifted his glass as if to toast his friend's success at a distance.

"To your success, Laci. My heartfelt congratulations."

He took a sip of the cognac before continuing to rummage through the package, to find a slab of chocolate, packages of tea and coffee, and, at the bottom, a letter. It read:

My dear Imre,

You always said that once we have missed our chance to be honorable, we can take our time bringing ourselves into disrepute. I put off writing this letter for so long that I thought one or two months more wouldn't make any difference! Please forgive me!

First and foremost, I would like to thank you again, from the bottom of my heart, for helping my family to escape from Hungary. It is only now we have read about Auschwitz and the other concentration camps that we can really grasp the fact that you saved them from certain death.

As the enclosed pen goes to show, my six years of work have born their fruit. The Birome pen is now available across the whole American continent, and since the end of the war it has come to Europe, and I am confident you will be able to get them in Hungary soon, too.

There is only one thing that matters: I would like to see you, without fail, and as soon as possible. I have sworn never to go back home; this is a rational, not an emotional, decision. Hungary is a poisoned land, where fascism can raise its ugly head again at any moment. You must come out to Argentina. Trust me, you won't regret it. I will send your plane ticket as soon as I receive your reply.

That's all for now. I am writing this in a hurry, because a friend of mine has offered to include it in a military postal shipment to Europe. I'm sending you a little bit of chocolate, coffee, and what have you, plus a couple of dollars; I hope you get these, too. Don't be shy—please tell me if there is anything you need, and I will send it.

With much love to you both,

Laci

Reminded of their former cooperation, Gellért felt obliged to show the Birome to Andor Goy. To no avail did he look for him in his old workshop at 9 Nádor Street: these premises had been taken over by the offices of state commissioner Zoltán Vas, and Goy had been apportioned a bomb-scarred shop in the corner of the Gresham Palace on Roosevelt Square. Together with some of his former colleagues, he had renovated it and turned it into a repair shop for pens and calculators. As Gellért entered, Goy stopped working and wiped his hand clean before greeting his visitor.

"I am glad to see you again, Mr. Gellért. How did you survive the times of darkness?"

"Don't ask me where I was, and don't ask what I did. If the dust ever settles on all this, we'll sit in for a coffee somewhere and discuss the whole thing. But let me come to the point. The reason I have come is to tell you that our brainchild, which once appeared to be stillborn, has come to life. *Voilà*: the ballpoint pen."

Gellért handed him the little box, and Goy took his time lifting the pen out of it.

"So this is it, plastic pen and plastic cap?"

Goy's visage revealed not emotions but the honest curiosity of a craftsman. First, he examined the pen from outside, seeing the name Birome on the clip, then looked at the writing tip itself.

"I developed this solution back in 1938."

He then unscrewed the body of the pen. He observed that the tip was connected to a thin reservoir pipe made of copper.

"This is not a new idea: this was already being used by the Evelco company before the war."

As a final test Goy took out his notebook and wrote a few words.

"It's only the ink that's new. Just the ink, nothing else. In every other respect I am in no doubt that this Birome pen is essentially identical to the ballpoint pen for which I have production rights for Hungary and fourteen other countries, according to our contract from 1938. That contract was signed in your name, too. Can I ask what plans Mr. Bíró has for the future?"

Gellért shrugged his shoulders.

"I can't speak for him. As far as I am concerned, I am willing to satisfy every aspect of the contract."

"I can see that from your presence here; I am glad that I didn't have to learn about these developments from a third party. Do you have a few more minutes to spare?"

"But of course."

"I would be pleased if you stayed for coffee."

As he stood next to his desk, Goy continued, contemplatively.

"Our relationship with Mr. Bíró reminds me of Jacob when he cheated Esau out of his birthright. Except that I haven't even got so much as the bowl of stew in return—that's what I would like to lay my claim to, in retrospect. Obviously there is no hope of an agreement that is friendly, but maybe we can make one that is peaceful, if nothing else."

"What exactly do you have in mind?"

"I don't know Mr. Bíró's current address, and in any case it would be unadvisable for me to get in touch with him personally. So I would be grateful if you could ask him to send me the formula for the dye. I hope he is reasonable enough not to deny this request."

Gellért wrote a letter to Bíró on the same day, conveying Goy's request. It was weeks before he received the reply from Buenos Aires. Bíró was unwilling to reveal the formula of the dye paste—it is worth noting that he makes no mention of the composition of the ink in his memoir, not even revealing when and how he achieved the final result. In his letter, he instead repeated his accusations against Goy:

"You should get everything, and Goy nothing! That man morally broke our contract when, in 1940, he wrote to me that I can go wherever I like, but that he, as an Aryan, will find me anywhere."

Gellért made every effort to skirt around the contents of the letter in a diplomatic manner, but Goy succeeded in getting the gist of it out of him.

"All I can say is that Mr. Bíró would be hard pressed to show anyone a letter in which I threatened him, because I never wrote such a thing, nor anything similar to it. As it happens, he is not the only one to cast such aspersions upon me: my business competitors have repeatedly accused me of having had extreme right-wing views during the war. They claim that the walls of my factory were decorated with swastikas, that I refused to supply goods to Jewish clients, that I sacked the Jewish director of the accounting machine division, that I signed my letters 'with Hungarian regards,' which was nothing less than a servile imitation of 'mit deutschen Gruss.'"

Gellért tried to interrupt, but Goy raised his hand to stop him.

"But these accusations are all untrue. I did not agree to the demand of the Füllhalterwerke to break off relations with non-Aryan business partners. During the war I hid a Jewish family—I would be only too happy to give you their name and address. The aspersions have all been examined by the People's Court, which instead of punishing me, bestowed me with praise. In their exact words, they wrote that the progress of democracy is helped by people who use their skills to support it, as I did. And now I would beg your pardon for interrupting you earlier—please go on."

Gellért did his best to jump to his friend's defense.

"I have no reason to question what you say; neither do I want to excuse Laci's actions. I am more interested in the explanations for his behavior. At the beginning of the war he clearly developed a persecution complex, for the simple reason that he really was being persecuted: they chased him halfway around the globe. I know very well what a struggle it was to save his family from disaster."

"Am I in any way to blame for all this? I gave him financial help, didn't I?"

"No one's saying that you are. All I'm trying to say is that a man can easily start to hallucinate in a situation like this and see all kinds of fantasies as if they were true. The other thing I can think of is that out there in Argentina he has become part of a cruel dog-eat-dog business world, and that it is difficult to shake off these bad habits."

Gellért had become exhausted by this catalog of explanations. Goy nodded to show that he had heard enough.

"To be honest with you, I'm not interested in Mr Bíró's emotional development. My only concern is that he has not sent me the formula for the ink I asked him for, despite the fact that it is his obligation to do so. If he wants to pick a fight, I'm game. I reserve the right to defend my interests through legal channels if I have to."

CHAPTER 23

T hough he had been working on it for many long years, the rigors of war meant that by 1947 Goy was still not ready with what he regarded to be his most important invention, a new typesetting machine. His experiments had succeeded only in draining his financial resources, and so he decided to put them on stand-by and concentrate the activities of his company on the production of the ballpoint pen. He looked through his warehouse for the materials he would need to begin; he found some one millimeter metal balls and a few dozen writing tips which had been worked on the lathe back in 1938. The production of the pen casing and cap was left to Gyula Kovács, a precision instruments maker, who persuaded Goy that using metal rather than plastic would offer a cheaper and more reliable solution. After extensive investigation they also found a suitable dye paste, the Victoria Blau Base material from the Ciba factory in Switzerland. Hungary's import company helped them to obtain adequate supplies to begin production.

The samples they made worked perfectly; wrapped in a machine-turned chrome case, the tip would write without leaving any gaps. All that remained to be done was to give the new product a name. Goy had always been bothered by the fact that Bíró had put his own name on the first ballpoint pen, so he now tried to give himself some sense of satisfaction by calling his version the "Gopen." He thought he had really hit the mark with this: apart from the English words "go" and "pen," it alluded to Goy's name, as well as to the Hungarian word golyó, meaning "ball."

Everything was in place for the Gopen to hit the market when an unforeseen development upset Goy's plans. The flea market on Teleki Square near Budapest's Keleti train station had already seen one or two traders selling imported Birome or later Birocco pens, but now a more dangerous source of competition had reared its head. Striking advertisements appeared in the press for a sensational new pen from America, the Akron, which, it was claimed, "worked with ink that sucks moisture out of the air." The product was released at an opportune moment, during the period before Christmas, and became an overnight success. Customers could obtain an Akron pen only after a complicated ordering process and queuing for hours, at a price of sixty-eight forints—at a time when housewives could do the weekend shopping for the whole family for ten to twenty forints.

These expensive pens ran out of ink after three or four days, and indignant purchasers who wanted their money back were offered only replacements. One scandal followed the next, and the shutters of the showroom in Ferenc Deák Street were rolled down not long after they had first been rolled up.

A couple of investigative journalists were quick to look into the Akron affair. They discovered that the pens had been produced not in America but in Huba Street in Budapest, under a license awarded to an obscure old widow by the name of Mrs. Zsigmond Herzfeld. The list of irregularities was topped by the failure to apply for permission from the patent owner.

Goy had every right to be worried that the failure of the Akron would be detrimental to the introduction of the Gopen, and so attempted to settle the matter as quickly as possible. He paid a visit to the owner of the franchise, a certain Imre György, and showed him the contract he had signed with Bíró, which gave him exclusive rights to the production and sale of the ballpoint pen in Hungary.

György tried to make the best of the situation by offering to join forces with Goy, but he refused, demanding that the Akron pen be removed from the market.

It was only threat of legal action that György was willing to cease operation, and he sold off his remaining shares at half its original price. Soon afterward he was arrested and sentenced to a lengthy prison term.

The memory of the Akron fiasco slowed the spread of the Gopen, but after a while it proved possible to build consumer confidence. Goy's business principle was a success: "The customer must never be cheated. Every product sold, if it is good quality, helps to sell another!"

The cramped production and distribution center soon proved incapable of satisfying increased demand. The Hungarian economy had not yet got over the troubles the war had inflicted, and difficulties continued to be present in the supply of raw materials and in international trade contacts. Goy's firm was unable to obtain the necessary copper tubes and the dye material produced by the Ciba factory. Uninterrupted production became more and more expensive, and despite his success on the sales front, Goy was still not making great financial returns. It was his good fortune to receive a visit from Gyula Mezei, director of the József Schuler Company, Hungary's largest pen manufacturer. The Schuler name brought back bad memories for Goy, however, to which he would make reference directly after Mezei introduced himself.

"I would like to ask you something, if you'd allow me. Where were you in June 1938?"

Mezei was a little surprised.

"As far as I can remember, I was on an instructional trip at the Hardmuth factory in Czechoslovakia. Is this relevant?"

"At that time I was making preparations for a business trip to Germany, where I wanted to give samples of the pen to companies there. But before leaving I thought I would give it a try at home. I approached the Schuler Company to ask if you were interested in the ballpoint pen. I met the director: he said simply that he didn't see any future in it."

Mezei tried to overcome the awkwardness of the situation with a jovial smile.

"The director in question happened to be my father-in-law. I hope you will give his son-in-law a chance to make up for his mistake."

"I'm the one who should apologize. But I felt I had to get this off my chest—his remark was hurtful at the time. Let's forget all about it. I am at your service."

Mezei offered to enter into a contractual agreement that Goy, after convincing himself of the technical excellence of the Schuler factory, accepted. He decided to continue his experiments on the typesetting machine, spending part of his time in his workshop on Roosevelt Square. At the Schuler factory he only maintained control of the production process itself, leaving the acquisition of raw materials and distribution to his contractual partners.

In his new role, Goy tried to revive his old German business contacts. He gave two Gopens to the director of the Rhine-Danube transportation company, who was on a visit to Hungary, to pass on to his former business partners at Deutsche Füllhalterwerke. They were quick to respond to this gesture, and were satisfied with the excellent quality of the samples.

They saw great commercial opportunities in the ballpoint pen, as they had before, but for the time being were not in a position to deal with it. The company's former director, Gerolf Zorn, had been exiled from the profession as a result of his Nazi past, and its direction had been taken over by a government commissioner. Zorn's brother promised that they would resume negotiations as soon as things at the company began to look up.

Goy was compensated by results at home: the sales graph was sloping upward steeply, and production was unable to keep up with demand. Goy attempted to persuade Mezei to transfer the capacity of the factory's metal works to the production of the pen, and to introduce a third shift, but the director was reluctant.

"If you had made this suggestion six months ago, I probably would have supported it. But right now I don't have a good feeling about it."

"You don't believe in the ballpoint pen any more?"

"Of course I do! I look out the window and see the wholesalers spending their nights here on the courtyard to get to their goods. But still, there is something hanging in the air."

Goy felt no danger. The fortunes of his typesetting machine invention were going well; it was even enjoying support from the government. The state-owned Corporation for the Sale of Inventions, founded under the aegis of the Inventions Office, covered the costs of the experiments, and later gave Goy a permanent position. His colleagues recommended him for one of Hungary's top national honors, the Kossuth Prize.

He could afford to buy himself a car, his first since the war. It was an old Ford Taurus. In late 1940s Hungary, a private automobile was considered a great luxury.

On October 13, 1949 Goy began his day as usual in the office of his workshop on Roosevelt Square. At nine o'clock in the morning he was sifting through the mail with his partner Kovalszky, when three men in leather coats came through the door. Goy looked up.

"Can I help you, gentlemen?"

"Are you the former owner of this company?"

Goy smiled.

"Excuse me, but what do you mean, 'former owner'? I am reminded of the joke about Schwarz visiting Mrs. Cohen to break the news of her husband's death, who asks at the door, 'Are you widow Cohen?' Except that I am not dead, yet."

"I can assure you that in a moment you won't be in the mood for jokes."

One of the men in leather coats took out an official document from his briefcase and placed it on the table.

"Read this. The Council of Ministers has ordered the nationalization of this workshop and has appointed me director, as of today. Please leave these premises immediately." He turned to Kovalszky, "Who are you, and in what capacity are you here?"

"My name is István Kovalszky, and I am the co-owner of the workshop."

"Then you must take your leave, too. I would ask you to give me all the keys, including those to the safe."

The newcomers, wanting to exploit the state of numbness all this would arouse in the owners, strove to conduct their business with the greatest haste. One of them completed an inventory of the valuables in the safe, while another gathered together the employees in the workshop and made an announcement about how the state of things had changed. The newly appointed director of the company repeated his request.

"Please put your coats on and leave immediately."

Goy gathered a few of his personal belongings. Lying on his desk was the first Gopen, the golden casing of which bore his wife's name, "Fernácska." He wanted to get this, too, but one of men in leather coats prevented him.

"Leave it, don't touch anything. We will complete our inventory, and if it turns out that anything was your personal property, you will get it back."

Kovalszky was unable to take anything apart from his work smock; and even that, only after the new company director checked its pockets. The two men stepped out into the street, pausing in front of the gates. Kovalszky bemoaned his fate.

"Do you mean to tell me that they can just throw us out of our own workshop?! That they can steal everything it has taken us decades of honest work to achieve?"

"It looks like anything is possible," said Goy, "but don't lose hope. They can't take our hands away from us, and so long as we can work, we will be able to make a living. Go home, get some rest. I'll call you in the morning. Maybe by then you won't see the world in such a bad light."

They went their separate ways. Goy set off for the neighboring Zrínyi Street, where his car was parked. He did not notice that his successor was following at his heels. He was about to open the door of his Ford Taurus, but the man stood in his way.

"The keys, please. This car has also been taken into state hands."

"You are wrong. This car is mine, I bought it out of my private earnings."

"If you had only read the warrant, you would have seen that the car has also been nationalized."

"I object! This is tyranny!"

The leather-coated man placed his hand on top of the car.

"If you make a scene, you could come to regret it. Should I call the police? You won't be done any injustice. If it turns out that this is your private property, you will have the right to reclaim the car, but you cannot take it now."

Meanwhile, the other two men in leather coats had appeared from around the corner, falling into line behind their leader. Aggression was not in Goy's nature, and he made no attempt to put up a struggle. As a sign of protest he pocketed his car keys, and started toward the Chain Bridge with slow, trembling paces. His daughters remember him, overcome by the shock, being brought home by strangers.

Goy was unaware of just how far the process of nationalization had come and hoped that Mezei would be able to give him some information. He collected his thoughts and made a call to the Shuler factory. The phone rang, but no one picked it up; the same was true for Mezei's home number. Goy continued his inquiries, getting in touch with the director's friend Kránitz, a paper trader in Dorottya Street, who told him that the Shuler factory had been taken over at more or less the same time as Goy & Kovalszky had been. Mezei had left Hungary with his family only a few hours earlier.

Production of the Gopen came to a halt when the smaller companies producing components were also nationalized. It seemed as if the fate of the ballpoint pen in Hungary had been sealed forever.

CHAPTER 24

All technical difficulties overcome, the Birome factory was producing at full capacity, churning out many millions of pens each month. With the owner's approval, it was decided that the key objective was to increase production. The time for new ideas and wide-scale solutions to problems was over; now the priority was to develop the best system of industrial organization—one that used time, materials, and energy in the most efficient way. This demanded leaders who were decisive and consistent, and who would use every method at their disposal to ensure that discipline was maintained in an environment prone to negligence.

Bíró was not one of these. "*Hola chicos!*" he would exclaim as he stepped into the factory every morning. He regarded every one of his workers as a friend he was happy to sit down and have coffee with. Neither was he in the mood to make any alterations to this way of doing things. However diplomatically it might be expressed, the dictatorial behavior of a superior was simply not in his nature. Bíró resigned from his position voluntarily, relinquishing the direction of the factory to an engineer contracted from Switzerland.

After years of continual distress and financial difficulties, Bíró wished to relax the tense pace of his life. Thanks to his wife's careful management of the family finances, he was in a position to do so.

Bíró retired to his home. The huge hibiscus bush that loomed high outside his bedroom window was blooming. The first thing he would set his eyes on every morning were its red flowers; he couldn't help but give a smile and say a little thank you to the hibiscus for bringing in another day.

Two or three times a week, Bíró would play golf at the course near his house, and in the evenings he would play chess or bridge with his friends. After a long lapse he was finally able to return to his favorite passion, painting. Bíró's works were well above the level customary for an amateur, and changed hands at high prices; some of them were purchased for inclusion in private collections. Journalists arriving to conduct interviews with him would often find him pondering in front of his canvas, turning the paintbrush with his fingers with the same composure that characterized all his actions.

"How is it that you combine a technical calling with a passion for art?" they would always ask.

"Painting was always my *violon d'Ingres*, the activity I could concentrate my enthusiasm on the most—if my work would allow it, of course. My feeling is that the inner forces that bring you to discover an invention or paint a painting are essentially the same. The same restlessness, the same excitement makes you achieve something."

"When did you learn to paint?"

"Never. I didn't even go to exhibitions very much. I was never in a position to take on the methods of the classical masters, because my paintings do not attempt to depict nature, only the creations of my imagination. Anyway, if it is obligatory to pursue formal studies, I cannot call myself an inventor, either, because I have not studied theoretical chemistry, physics, or mechanics."

Bíró was unable to pursue his gentle pastimes for long, however; his colleagues would not let him live the quiet life. Meyne, his feet always on the ground, surveyed the opportunities offered by the South American market. As he saw it, the smaller countries would not be in a position to set up their own production facilities, and so it would be worth establishing a company to distribute the pen in the region. In his heart of hearts, Bíró would rather have rejected the offer, but his conscience prevented him: both his brother and Meyne had been squeezed to the sidelines of the Birome factory and its associated institutions, and the new Bíró-Meyne-Bíró Commercial and Industrial Company would offer them suitable employment.

Meyne continued to put appearances first. To celebrate the founding of the company he sent a pen as a gift to the president of Argentina, the federal commissioner for the Buenos Aires region, and other dignitaries. He published their thank you letters in the newspapers. For the company headquarters he rented out two whole floors in one of the office buildings downtown, from where he would "convert" the countries of South America to the ballpoint pen. The new enterprise soon delivered on its promises.

Bíró could most strongly feel the easing of the political situation in Hungary that ensued after 1953 in the hugely increased amount of mail he began to receive from his homeland. Newspaper articles appeared in Budapest that referred to him as a "dollar-millionaire inventor." These led to a swathe of letters begging for money or support for applications to emigrate to Argentina. While all of those who made such pleas introduced themselves as his intimate friends, or at least close acquaintances, many of their names were completely new to him. To begin with, Bíró was only too happy to help, then began to tire of the endless stream of requests, and the letters hit the wastepaper basket without a reply.

One day he was delighted to find a letter from Imre Gellért amid the mail pile. His former partner had written to say that he would be happy to accept Bíró's earlier invitation and, making the most of the theater's Christmas break, would bring his wife with him.

Buenos Aires was already making its seasonal preparations as the DC-6 plane made its descent onto the tarmac; the airport loudspeakers were ringing with Christmas carols. Most of its passengers were Italians paying a visit to their relatives. The Gellérts deliberately lagged behind their noisy companions, and were the last to pass through customs.

It was more than thirty-six hours since the couple had taken off from Budapest: the plane had landed in Zurich and Geneva, made a refueling stop in Lisbon, and stopped also in Dakar, Recifé, Rio de Janeiro, and Montevideo. Yet to Bíró it was as if they had just stepped out of their apartment in Budapest. Before disembarking they had brushed themselves up and they radiated the aroma of fresh perfume. Gellért's wife handed Bíró a small package.

"*Beigli* rolls—one poppy-seed and one walnut. I baked them for you for Christmas, Laci, you have to have a taste!"

Bíró kissed her hand, then hurriedly packed the pastry away. He did not have the heart to tell her that in Argentina poppy seeds were sold only in pharmacies, as they were considered a narcotic, and it was illegal to bring them into the country. He gave Gellért a resounding hug, then stepped back to look at the suit he was wearing.

"Is this the same suit you had made at Staub's in Semmelweis Street back then in Budapest?"

"That's right, from the first money we got for the pen. It's still as good as new. I paid a thousand pengős for it, but it was money well spent."

Outside the airport Meyne was waiting for the guests with his car. Gellért extended his hand.

"The last time we met was in Paris, in 1940, wasn't it?"

"Exactly right," said Gellért, slapping his forehead in disbelief.

"What do you say?" said Meyne. "We've got a whole little army here for you."

It was only then that Bíró realized how much Gellért and Meyne were alike. Whatever would be taking place in the world around them, they would move around at ease, always finding enough strength to begin new initiatives; they would never make martyrs of themselves, but enjoy the spice of life with good humor and elegance. For all the affection he held for them both, he could not help feeling a little envious. The only way he could console himself was to remember that the style both Gellért and Meyne employed would never lead them to great achievements: he who wants to get far into the jungle has to clear himself a narrower path.

It was a warm spring evening, and Elza proceeded to set the table on the terrace. It was in the conversational style of his youth that Bíró invited the others to come.

"I hope you will pardon the fact that my wife is doing the serving, but the fact is that my butlers and my chauffeurs are playing each other at soccer tonight."

The Argentine cuisine was accompanied by Hungarian wine and followed by a taste of the walnut and poppy-seed rolls. It was almost midnight when they got up from the table, but Gellért was still fresh and stayed with Bíró on the terrace for a final cigarette.

"Do you want a Korona? That's what you used to smoke back home, if I remember right."

"Thanks. What are your plans, Imre?"

"What do you mean?"

"Haven't you thought about staying here? You always said that you never wanted to leave your mother alone. She's passed away now, poor thing, so that can't hold you back any more. You haven't had a chance to look around yet, so please take my word for it: Argentina is like a little Europe across the ocean—and the better face of Europe at that. Buenos Aires is more akin to Paris than to that bigoted Madrid. Argentina is the country that has made the most progress in intellectual freedom. It has learned to have a laugh at priests, it holds its future in its own hands, and it takes a critical view of beliefs that yesterday were still holy."

"And what are the people like?"

"It is said that the Argentines don't like anyone, least of all the Argentines. My experience is that they will accept anyone."

"That's all very well, but what could I do with myself out here?"

"You would have no trouble finding your feet. This is a young country. The intelligentsia is still a paper-thin sector of society—real talent is rare, but all the better appreciated for that. It is easy to find investment for enterprise, and whatever you produce, it is no problem selling it, the market is so big and open."

"All right, let's say I don't have any trouble acclimatising to things, but what would happen to Olly? A Hungarian actress is only of any use within the borders of Hungary, where the audience can understand what she is saying."

"She would find her place, too. If we don't watch out, there will be more Hungarian actors working in South America than back in Hungary: Zita Szeleczky, Pál Vaszary, Miklós Hajmássy, Antal Páger, and his wife. There's a permanent Hungarian theater here—they'd be more than happy to welcome Olly. I hardly have to tell you that you can count on my help in everything."

"Thank you, Laci."

"My motives are entirely selfish. It would be good for me if you were close by."

The two friends fell silent for a moment. The Southern Cross constellation dimmed in the sky above them, and the bright red flowers of the hibiscus slowly began to make themselves visible in the gloom.

"I'm not trying to steamroller you, Imre. Why don't we agree that you'll give it a try? Join the company, at Meyne's side; the only thing you have to do is learn Spanish, and watch them as they work. If you don't like what you see, you can quit whenever you like."

Gellért accepted Bíró's offer. He was given a desk in one of the offices, but he preferred to spend his time down in the factory. He would read through brochures and would often take a ballpoint pen out of its box, take it to pieces, and study the components for hours. He ordered up a variety of small containers from the warehouse, pouring mixtures of perfumes into them. One day he paid an unexpected visit to Bíró.

"Laci, I have an idea. Might it be possible to apply this ball idea to other products?"

"Can you be more specific?"

"In my opinion, fashion items are the only ones worth dealing with. A cologne bottle, for example, that doesn't work with a vaporizer, but spreads the scent on the skin with a rolling ball."

Gellért showed him a hastily drawn technical outline, which Bíró looked at thoroughly before nodding.

"This is a great idea; I'll buy it. How do you want to receive your royalties: cash or share of rights?"

"Cash, only cash. I've no idea how long I'll live and I don't want strangers to inherit my money."

"That's up to you, though you'd do better with a share of the rights."

Bíró wrote out a check for two thousand dollars. Gellért looked at it suspiciously.

"Are you sure this isn't a covert act of charity?"

"Don't be ridiculous!"

"Then that's all right." Gellért stood up straight. "My dear Laci, I'm so sorry, but I quit, this very minute. Thank you for everything you've done for me."

Bíró didn't want to believe his ears.

"You want to go now, at the very beginning of your success?"

"Argentina doesn't bring me success. The only success for a Budapest lad like me is if they are talking about him in the cafés on the Grand Boulevard back home."

Bíró did his best to dissuade Gellért of his intentions, but to no avail. Only a few weeks later they said their goodbyes at Buenos Aires airport.

"Come again, Imre!" said Bíró. "As the Germans say: *einmal-keinmal!*"

"It's you who should come home, while I am still alive, if you can. God bless!"

CHAPTER 25

M eyne and Bíró's older brother sensed real business opportunities in Gellért's idea. They set out the technical
requirements in painstaking detail. They wanted to sell
the roller-ball perfume in a case similar to that of lipstick and fill
it with the highest-quality French perfume. The composition of the
substance was left to the Lengyel brothers, highly qualified professionals who had recently immigrated from Budapest.

The entrepreneurs thought it best to begin distribution of the
product not on the conservatively minded Argentinean market, but
in the more fashion-conscious United States, and so they decided to
open an office in New York.

The day-to-day activities of the company directors tied them to
the headquarters in Buenos Aires, and so they found it necessary to
find a suitable director for the New York branch. György Bíró's wife
recommended her brother, Emil Rapaport, who lived there, for the
job. Before coming to a decision, however, Bíró wanted to make his
acquaintance, and paid him a visit in the Big Apple.

Rapaport, a short young man with a large head and a rakish mous-
tache, began his career as an actor before shifting to the world of business. The references he made and the names he dropped during their
conversation created the impression that he had a first-rate network
of contacts. Bíró saw through his schemes, but neither could he come
up with any tangible reason to reject him nor did he wish to enter into
an argument with his sister-in-law; and so, not without reservations,
he gave Rapaport the backing he required. Bíró gave the new director

of the company the samples of the product he had himself created, on the basis of which the mass production of the components could. begin. Bíró offered to provide continual professional supervision, but Rapaport rejected this with a gentle smile.

"I've solved bigger problems than this in my time."

A hundred-thousand-dollar publicity campaign accompanied the introduction of the product, with the cooperation of the Garant Agency, one of the leading names in the advertising industry. The first results were quick to come. Rapaport announced in the press that among the world-famous companies applying for production rights was the Cartier group, which wanted to make its roller-ball perfume holder out of eighteen-carat gold. The youthful director responded to this success by using the Parker slogan:

"Make something better, and people will buy it."

Meyne declared victory, but Bíró chose instead to wait. During his career he had learned that unsuccessful battles begin with the same gunshots as do victorious ones. His misgivings quickly proved well-founded; the promising start was soon strangled by an unexpected technical disaster. Rapaport, inexperienced in industrial matters, had bought badly-sized metal components from his suppliers, and the roller-ball perfumes leaked their contents in the same way that the early versions of the ballpoint pen had spilled their ink. Goods made their way back to the warehouse in their entirety, and, having reneged on delivery contracts, the company was forced to make huge compensation payments.

There was no hope of making a new, perfectly working model of the product. Consumer trust had been lost, and they did not have the financial resources to start the kind of publicity campaign with which it might have been regained. Yet, Rapaport remained unmoved by the whole affair.

"The only people who don't make mistakes are those who never do any work. I will probably return to my acting career," he told the press.

The final balance showed a loss of two hundred thousand dollars, which not only wiped out the entire capital of the Bíró-Meyne-Bíró

Company, but the restructuring also necessitated the sale of two-thirds of the stake the company had in Birome.

Meyne said his goodbyes to the Bíró brothers and took up a position at the El Globo advertising company. Soon afterward he married the widowed owner, Maria Fulgencia. He would occasionally make a visit to his former workplace, in even more elegant attire than before, driving the latest Mercedes.

"If someone wants a Hussar officer, they should be prepared to pay for one!"

The fiasco did not take away Bíró's appetite for experimenting. New ideas would constantly form in his mind. His imagination was most occupied with a wristwatch-shaped gadget for continuously measuring the blood pressure of its wearer. The inventor would have set to work on research work were it not for the fact that the failure of the roller-ball perfume had robbed him of all his financial reserves. He had no option but to turn to the Lángs for assistance. Mrs. Láng was more than happy to help, asking Bíró to deposit his remaining Birome stock with her as collateral.

"This is just a formality, Laci," she said, "as if you were taking your money out of one pocket and putting it in the other."

Bíró did not sense any particular risk in this matter. The stock was in his name, and thus could not be transferred to anyone else without his prior knowledge and agreement. He was in no mood to part with it, however, for under Henry George Martin's expert and fortuitous leadership it was consistently increasing in value.

A few months later the experiments demanded that Bíró travel to Switzerland. By then he had collected together enough money to pay back Mrs. Láng's loan. Before leaving he paid her a visit, but she refused his offer to pay off the debt.

"You are going to a foreign land, Laci! Life is always more expensive there, not to mention surprise expenditures. Keep the money for now; the loan is free of any interest."

The inventor was touched.

"I always knew you were a real friend, Mária."

"In return I would ask for a small favor. As you know, there are some questions in which I do not see eye to eye with Martin's group, but the only way I can enforce my recommendations is if I use your voting rights in your absence. I would be very glad if you would give your assent."

The following day Mrs. Láng added a further request. She asked Bíró, as a formality, to sign a release form for his shares—failing to mention that this effectively meant a transfer of ownership. Alarm bells were already ringing in his head—the memory of the earlier manipulations of the Lángs was still very much with him—but he felt indebted to Mrs. Láng for the extension of the credit, was unable to say no, and signed the "formal" release of his shares.

Bíró was later to recognize his mistake as follows: "I am obliged to admit that in this matter, which later led to one of the greatest disappointments in my life, I was thoughtless and did not display adequate scruples."

He tried to underplay the significance of the issue, convincing himself that the fate of his stock would pale into insignificance compared to the expected results from his trip to Switzerland; his further initiatives promised greater and greater returns.

Bíró's stay in Switzerland was longer than planned. From time to time he would make a trip back to Buenos Aires. On these occasions he would try to settle his debt, in the hope of getting his shares back. Mrs. Láng continued to dismiss all such attempts.

"Laci, these shares are too valuable to be in your possession. I know your spontaneous way of doing things. If once you need some money, you would be capable of trading in everything for a couple of beans."

Bíró thought he could allow himself a sarcastic comment.

"So you think that knives, forks, and scissors are to be kept away from children?"

"Trust me, they are better off with me; I will look after them for Elza and Mariann. One day you will be grateful to me for it."

After a year the inventor returned home for good. He was greeted with the news that Láng had died, and that his wife had been hospitalized with serious heart problems. It was only really out of politeness that Bíró paid her a visit. Given what had happened, he thought it in bad taste to bring up the matter of the shares all over again. But Mária mentioned it out of her own accord.

"I have to be here for another two or three weeks, but as soon as they let me out, the first thing I will do is give you your shares back. But if, God forbid, something might happen to me, you will not be done any injustice. I have stipulated in my will that this group of shares rightly belongs to you and should be returned to you."

The following day, however, Mrs. Láng had a sudden heart attack and her suffering came to an end. On hearing the news, Bíró said the customary Jewish prayer of mourning under his breath.

"With praise, Our Lord, you are the Greatest Judge!"

After the funeral, the family's lawyer—who had made a few pennies of his own from the invention years earlier—read out Mrs. Láng's will. She recognized Harry Kleinlein, her son from her first marriage, as her primary heir; there was not even a mention of Bíró.

In Switzerland, Bíró had often met with Kleinlein, who had been studying in Zurich. The disabled young man, perhaps as a way to express his dissatisfaction with the world, held anarchist and anti-capitalist views. It seemed possible that he would remain loyal to his beliefs, and Bíró hoped that this might increase the chances of him wanting to right his mother's wrongs. But when he came to pay him a visit, he found a newborn capitalist who refused so much as to touch upon the question of the ownership of the shares.

"You've changed your tune!" said Bíró, reproachfully.

"That may be, but every vegetarian changes his ways when he inherits a sausage factory."

And so it came to pass that the inventor who conducted the thousands of experiments needed to perfect the ballpoint pen ended up without a penny of stock in the factory where they had taken place.

CHAPTER 26

After the appropriation of his factory, Goy became an employee of the succession of state companies that took each other's place in turn, working on the perfection of his invention, the Typopress typesetting machine.

Unable to support the four other members of his family from his salary, he was forced to look for moonlighting jobs. Many of the other residents in his area of Budapest were labeled "class-enemies." They banded together and formed a cooperative with Ede Eckert, a former freight executive, as its leader. They produced bedcovers, thick as carpets, for sale in the provinces. They painted, steamed, and acidified the material in a basement workshop. The finished bedcovers were hemmed by Ferdinanda, Goy's wife, with a Pfaff sewing machine. Whenever a larger bundle was ready, Goy would carry it down on his back to the cooperative's premises on Fodor Street.

In accordance with the communist government's Decree IV of 1952, Goy's house was also taken into state hands; the fact that he was working on an invention that he was offering to the Hungarian state did not grant him impunity.

Following a long list of setbacks—which would themselves fill a separate volume—Goy had, by 1958, made enough progress to be sent as a member of a four-member delegation to West Germany. His task was to present the Typopress to the Adler factory in Frankfurt.

The delegation left Budapest the afternoon of November 22 on the express train to Vienna, there catching a night train for the longer journey to Frankfurt. Goy shared a sleeping compartment with an engineer by the name of István Halász, a department head at the state-owned Licencia Corporation for the Sale of Inventions. Halász was himself an inventor, and was going to hold meetings with the Faber-Castell company about a new kind of slide rule.

It was the first time in many years that both men had crossed the border to the West, and they could not sleep for their excitement. Before the night train left Vienna, Halász jumped off to get some sausages and rolls from the station snack bar, along with several bottles of beer. Back aboard, he and Goy sat opposite one another on the edges of their beds, nibbling the spoils from paper plates in their laps. A long whistle rang out, whereupon the train gave a start and left the station. Goy stared out the window at the railway warehouses they were leaving behind.

"Do you know your way around these parts?" asked Halász, his mouth full.

"Before the war I would take this train two or three times a year. I made trips to the Wanderer typewriter factory—I was their representative in Budapest. In the meantime I would spend my time looking for partners for the ballpoint pen."

"What does the ballpoint pen have to do with you?"

"Back in 1938 I signed a contract with László Bíró. His name is familiar, I presume."

"Of course, the world-famous inventor."

"In those days he wasn't world-famous yet, and indeed he was on his last dime. I financed his research and made a contribution to it, too; in return I was granted the right to produce the pen in Hungary and another ten European countries. If you don't believe me, come to my house sometime and I will show it to you."

Halász was about to take a sip of his beer, but stopped short.

"Are those rights still yours?"

"Yes."

"I don't understand you, Andor. With such an opportunity in your hands, why didn't you leave for the West? You could have made millions."

"As it happens, I would have had the chance. At the start of

November 1956 a car was waiting outside our house. I knew the driver—all I would have had to do was get in with my family. But I wanted to make my career at home in Hungary. I had faith in my invention, the typesetting machine, as I still do now. I know only too well how lead has ruined the lives of printers; this machine does not need even an ounce of lead."

Day was already breaking when the train pulled into Frankfurt's main train station. The Hungarian delegation immediately set off for the site of the technical presentation. It proved a success. The directors of the Adler factory were impressed by the machine; as far as they were concerned, they were happy to buy the rights at once. They were not in a position to make the decision there and then, however, for their company had recently been purchased by the Grundig group. They asked the Hungarians to make a further trip to Nuremberg to register at the central patent office. This change to their plans was not so simple to arrange, as the movements of the Hungarian mission were under strict state supervision. The delegation's minder—Mrs. Béla Verő, a state security agent disguised as a trade executive—had to send a special request to Budapest, and it was only when they received a positive response that they could continue their journey. The invention went down very well with the specialists at Grundig, too, who asked them to leave it for them to carry out the necessary tests.

Having completed the task in hand, Goy retired to his hotel room, prepared a report on his trip, and had a rest. He was woken abruptly by a call from his fellow delegate István Halász, whose voice was full of excitement.

"Sorry to bother you, Andor, old man. I'm calling from the Faber-Castell factory—the folks here would really like to meet you."

"Meet me? Why?"

"I bragged that I am sharing a room with Andor Goy, the inventor of the ballpoint pen."

"Are you crazy? I never said that."

"You own the rights to the production of the pen, so it's not such an exaggeration. Herr Sattmann, the director of the factory, still remembers you from your visit in 1938. You can't possibly refuse their invitation."

With Goy still hesitating, Halász began to beg.

"You would be doing me a great favor as well—the prospects for my slide rule will improve if you come. They are sending a car to get you right now."

And with that, to quell any further resistance, Halász put down the receiver.

Faced with the inevitable, Goy began to get dressed. As he straightened his tie in front of the mirror, he was reminded of the meeting twenty years earlier. He could not help thinking back to Sattmann's snub, when he had even questioned the presented ball-point pen's status as a stand-alone invention, and to the way he had returned the pen with not so much as a "With Compliments" slip. At the time Goy had managed to take this treatment in his stride, but now, at a riper age, he thought it very unfair. The telephone rang once more. Goy determined that if it were Halász again, he would turn down the invitation, but it was the porter's voice at the other end of the line.

"Mr. Goy? Please make your way down to the lobby; the Faber-Castell car is waiting for you outside."

There was no way Goy could avoid Mrs. Verő, who was stationed on a couch in the lobby.

"Where are you going, Comrade Goy?" her authoritarian voice boomed.

"To look for a bookshop. I would like to buy Marx's *Das Kapital* in the original German."

"Very good. I can accompany you if you'd like."

"You're very kind, but you must be tired after such a hard day."

"Tired indeed. I have to keep an eye on everything."

Goy was afraid that the "commercial executive" would watch him leave, and so he walked on a few yards before getting into the waiting Mercedes. The drive calmed his emotions; he watched the city's preparations for the festive season with considerable delight. The Christmas lights, stretched between the houses, filled the streets with a homey glow. The old city center had been renewed and the streets were lined with modern buildings, with only the façade of the Faber-Castell factory, looking like a medieval castle, harking back to earlier times.

The chauffeur took Goy directly to the meeting room, where Halász and some of the directors of the company were waiting for him. Sattmann rushed forward to welcome his guest, clasping Goy's hand with both palms.

"You don't look a day older than all those years ago."

"Age is only important while you're getting old. Now that I *am* old, I feel like I were twenty. You could easily pretend to be ten years younger yourself."

"This meeting calls for a celebration!" said Sattmann. He gestured to a waiter, who pushed a trolley into the room, full of cognac, champagne, and sandwiches. He lifted his glass to propose a toast.

"It's taken me a while, I'll admit, but I'd finally like to apologize for so stupidly rejecting your offer all those years ago. If only I'd had more sense, the ballpoint pen would have started its march to victory from here in Nuremberg!"

"Don't be so hard on yourself! Back then I got the same response from Pelikan in Hannover and Mont Blanc in Hamburg."

"I find no comfort in that. I would like to make up for my mistake, if that it is still possible."

Goy returned his gaze, his expression quizzical.

"What do you mean by that, Herr Sattmann?"

"I was talking to Mr. Halász before you arrived. He tells me that the contract you signed with László Bíró is still binding. Is this true? And more importantly, can you prove it?"

"Yes, all the papers pertaining to this are at my disposal. During the siege of Budapest, a number of bombs hit our house and much of our furniture, dishware, and clothes disappeared in the ensuing chaos, but it seems no one was interested in these documents."

"I'm very glad to hear that. If that is really how things stand, the Faber-Castell company will make you an offer. I'll leave it to the head of our legal department to give you the details. Dr. Czirnich, please!"

"First of all, we would ask you to send us a photocopy of every document that may act as evidence in the event of a court case."

"What court case do you mean?"

"The case you would bring against the legal successor to László Bíró, the Biropatente Company of Switzerland, in the defense of the rights granted to you by the original contract."

Goy was lost for words. He let the lawyer continue.

"Please send us the facsimiles as soon as they are ready. We will examine them, and if we think that there is a good chance of the case being successful, we will help you begin proceedings. We can discuss the precise legal strategy in due course. Probably the easiest way to proceed would be for you to move to Nuremberg with your family. You would be given a suitably furnished apartment, and we would create a job for you in our factory. This is where you would live until the court passes its judgment."

"I told you it would be worth you coming, didn't I?" interjected Halász, putting his hand on Goy's shoulder. But the dazed inventor did not respond to this overly familial gesture.

"This is not the sort of offer I can give an answer to out of hand. But the biggest question I have is this: what are you hoping to get out of this case? I assume it is not altruism that is motivating you to help me. I have no need for your charity."

It was Sattmann who provided the response.

"Yes, that is a fair question, but one with a very simple answer. I'm sure you are aware that Faber-Castell produces ballpoint pens; millions of them come off our production line each year. But you can't possibly know how much we pay to the Biropatente Company in royalties. They make the most of their monopoly, setting a fee that is much higher than anything seen before in Germany: thirty-five hundredths of a cent per unit. Last year alone we transferred more than two million German marks to their bank account, and this year that will go up to three million. This is an intolerable state of affairs."

"What has this to do with me?"

"If the court were to rule that the contract you signed with Bíró is still valid, you would have the right to demand royalties from now on. I am convinced that we would be able to arrive at a more reasonable arrangement, under which both sides would do well. This possibility is worth investing a small amount of money in."

"Can I try to put that in words a layman can understand? You are offering to foot the legal bill for the case?"

"We are, but officially you would be the plaintiff, as the person whose interests are directly affected. There is no question that the procedure would have to be initiated from Germany; it is almost impossible to begin successful litigation from behind the Iron Curtain. As I said, this is only possible if you move here, at least for the time being. That is everything I can tell you right now. I hardly need to tell you that this matter demands the utmost discretion. When can I expect your decision?"

"I can't make any rash promises. Things in Hungary don't work as they do here, I have to go through a few legal twists and turns."

"I'm afraid that is beyond our sphere of influence. It is the bee's job to fly into the hive, as the German saying goes." Sattmann stood up. "It was a delight to see you again, and I look forward to hearing from you, Mr. Goy. The chauffeur will take you back to your hotel."

On reaching downtown, the two Hungarians asked the driver to stop outside a bar, thanked him, and went inside for a beer. Halász lifted up his glass.

"Congratulations, Andor, old man. A great success, or at least a good chance of one."

"You're very kind, but I am still dazed by the whole thing. What would you do if you were me?"

"I don't think you can get away with only feathering your own pillow with this affair. They say the authorities in Hungary don't read through the mail any more, but I don't believe that. But let's assume fortune favors you and you manage to get the photocopies out to Nuremberg. Sooner or later they will find out, and put you in jail for currency offences. They'll kick your kids out of school. You have no choice but to find the person best suited to your needs."

"This is your home turf. Who should I turn to?"

"Do you know Géza László, my boss at the Licencia Corporation?"

"I've met him a couple of times, back in the days when he worked at his brother-in-law's company, the Friss and Guido Shroud Factory. They bought their typewriters from me."

"I'm sure he remembers you, too. Try to talk to him before you do anything else."

Goy was almost at a loss for words.

"This whole thing is so complicated—maybe it would be better not to get into it in the first place. It looks like the typesetting machine will become a success in Germany, so maybe that's what I should spend my time on."

"That's a decision only you can make, my friend. But in my view you have the trump card in your hand, and that's the ballpoint."

It was past midnight when they returned to the hotel, but Mrs. Verő was still there in the lobby.

"Did they have what you were looking for, Mr. Goy?"

For a moment, Goy thought she had caught on to his dealings at Faber-Castell, before remembering that she was referring to his alleged trip to the bookshop.

"They were out of stock, as it happens, but put it on order and promised to send it by mail."

CHAPTER 27

B ad news awaited Goy on his return to Budapest. The National Typopress Company, where he worked, and which dealt with the development of the typesetting machine, had given him notice, saying that they had put a halt to the research. They offered him the opportunity to continue his work at a different factory, but as Goy had previously had bad experiences with this other institution, he declined. He could not help thinking that fate had made the decision between the two great opportunities in his life on his behalf.

From then on Goy concentrated his attentions on the ballpoint. He collected the various documents he needed, and, taking Halász's advice, took them to Géza László, director of the Licencia Corporation. He recounted his dealings with the leaders of Faber-Castell, and asked for permission to send them photocopies of the documents. László was reluctant to commit himself to anything.

"This is a highly complicated issue; we have to discuss it with our legal experts. I would ask you to leave these papers with us and we will contact you when there is any progress."

Goy handed over the dossier.

"How soon could I hope to get an answer? I don't want to press you, but the fact is I'm out of a job and a livelihood, and I have to support the other four members of my family."

"I was told that you have another invention, a certain typesetting machine—is that not going so well?"

"No, it's stuck in a rut for the time being."

"I see. I cannot promise to speed things up, I'm afraid. Wouldn't you be interested in turning your attention back to the ballpoint pen in the meantime?"

"How would I get a license to trade?"

"I didn't mean producing the pens yourself—you could join a cooperative that manufactures them. I'm sure you could be of mutual benefit to one another."

László introduced Goy to Endre Káldori, director of the Hungarian Pen Manufacturers' Cooperative (the legal predecessor to ICO, which is today Hungary and Central Europe's largest ballpoint pen manufacturer). Goy greeted him as an acquaintance, remembering him from the days when Káldori was still learning the fountain-pen repairman's trade at the workshop of his uncle, Zoltán Schwarcz, on Liberty Square.

"You have always been my role-model," Káldori said. "There is a Gopen in my desk drawer to this day. The only reason I didn't make contact with you before is that I was afraid you would think me impertinent. I am very happy that Comrade László has introduced us like this."

Goy paid an initial visit to the cooperative's premises in downtown Budapest. His jaw dropped when he saw that production was carried out on a village industry scale: the primitive lathe was only used to make writing tips, while all the other components were imported. Káldori began to make excuses.

"Don't be influenced by what you see here, Andor, my old friend. Our new headquarters on the city outskirts will be ready by the start of next year, and we are already beginning to buy the latest machines for it."

Goy was no youngster when he became a member of the cooperative on January 15, 1959. Three years later, they signed a contract transferring the production patent, in return for five percent of the factory price of the pens made. In the meantime, Goy received a monthly salary of 1,900 forints, hardly more than the average for a skilled worker. Initially he did not even have a desk of his own, but he was promised his own office with a telephone in the new facility.

The agreement also included a nebulous closing paragraph, stating that it would only come into effect after the cooperative's national head office had given its approval. Goy did not attach too much significance to this, just as he was not perturbed by his relatively low starting salary. He was convinced that the case against the Biropatente Company would decide his fate.

Káldori made use of Goy's expertise, asking him to help decide what machines to purchase for the making of the pens. Having weighed up their options, they decided to buy used equipment from the Schneider factory in Tennenbronn, West Germany. They advised the ministry of their intentions, but it was six months before they were given work passports and the hard currency required for the purchase. Before leaving, Goy paid Géza László a visit to ask how the case was developing and whether his presence was required.

"You can feel free to travel for the time being," said the director. "Maybe by the time you return a decision will have been made."

In light of this continual procrastination, Goy decided to take the affair into his own hands and not leave it to the whim of the authorities. He dug out one of the old Gopens, filled it with a Swiss dye material, and put it in his pocket. He composed a twenty-four page description of the ballpoint pen in German, in which he summarized the history of the pen in Hungary, making special note of the contract signed with Bíró. He put the typescript in his suitcase, well aware he was committing an illegal act; he was only allowed to take documents with him if they were first reviewed and approved by the appropriate government agency. If the customs guards found it on him, they would send him back. Goy was more than a little relieved when their inspection was over. He recalled the old saying:

"Sloppiness is the best answer to dictatorship!"

In his memoir, Goy writes that he was positively dumbfounded by what he saw at the Schneider factory. Two dozen so-called "automatic plate machines" stood in an enormous hall in two parallel rows. They turned cut-to-size pieces of copper into writing tips with almost nothing in the way of human involvement. The pace of technical progress was always increasing. Earlier, one writing tip had been turned on the lathe every four minutes, as in Hungary; now each of the machines produced thirty a minute. The newest ones were capable of turning out sixty to eighty.

"None of this is adequate to satisfy demand," Goy's German hosts told him. It was clear to him once again just how much wealth even the most modest of royalties would bring him.

At the end of their tour, the Hungarian visitors decided to buy two or three of the automatic plate machines that were about to be replaced. The visit was followed by lunch in the factory cafeteria. The heavily diabetic Káldori had to turn down the invitation; he gave himself an injection of insulin and had a rest in the guest room. His only potential witness now well out of earshot, Goy did not have to worry about the Hungarian authorities keeping abreast of developments, and made a tactical maneuver.

At the conversation after lunch, he first showed the Gopen to his hosts. Though produced more than ten years earlier, the pen still wrote perfectly.

"If you could manufacture such a fine product, why did you stop?" a company official asked him.

"My firm was nationalized in 1949. They came and took everything."

"But why didn't you call for the police?"

Thinking it best not to continue explaining, Goy instead changed the subject. He made reference to the contract he had signed with Bíró, saying that he had exclusive rights to the production and distribution of the pen in a number of European states—including Germany. When his hosts expressed doubts about this claim, Goy opened his briefcase.

"Gentlemen, if I were to start the story from the beginning, it would last for hours," he said, taking out the document he had written. "It would be easier if I gave you these few pieces of paper, in which you will find every detail of the affair and the role I played in it. If you find it of interest, I authorize you to make copies of it for yourselves. I should add that I would be grateful if Mr. Káldori doesn't hear about our conversation. As part of his role here, he has a duty to report certain things."

Schneider chief executive Christian Schneider and his colleagues read through Goy's memorandum then and there. Just as the owners of Faber-Castell had been, they were excited by the possibility of not having to pay unreasonably high royalties to Biropatente in Switzerland but instead coming to an agreement with Goy, the new owner of the rights.

In their opinion the progress of this issue would have an impact on the whole German pen industry, and they did their best to attract publicity to it. They extended the stay of their Hungarian guests by a few days, during which time they introduced Goy to Robert Göller, the owner and editor-in-chief of the trade paper *Fachzeitung Papier*. The journalist conducted an exclusive interview with the visiting engineer, during which he learned of the existence of the manuscript. Having read it, he asked to be allowed to publish it.

After this, there was no way Goy could avoid informing Káldori of developments, though he stopped short of telling him that he had brought a completed document out of Hungary with him.

"The fact is I was getting bored of telling the same story again and again, so I wrote a summary of the ballpoint story in German while I was here. The press have asked if they can publish it. What do you think, can I give it to them? I think it would strengthen our professional prestige here."

Káldori was too much of an old hand really to be fooled by this, but he restrained himself from giving rise to a conflict.

"I'm not familiar with the text, and my German is not good enough for me to read it. There is just one question I would ask. Is it in our best interests? Does it put the socialist system in a bad light in any way?"

"There is no way that it can because the story begins in 1938, and lasts until the beginning of the war. Actually I think that a lot of the ideas in this document are yours as well as mine—that is why I would like to include your name as coauthor. It doesn't do any harm if they know your name in Germany. It can prove useful when doing deals."

"My dear Andor, I'd be honored."

Goy and Káldori were still on their way back to Hungary when the first instalment of the story was published, entitled *Aber die Kugel, die rollt—But the Ball Must Roll On.*

CHAPTER 28

The official approval procedure continued to make no progress. Goy was afraid that those responsible had already rejected his request, and were not informing him for tactical reasons or simply out of convenience. A few days after returning home, he paid Géza László a visit to find out for himself.

"I am glad to see you!" said the director. "Káldori has already told me how successful your trip proved to be. I hear that you have even appeared in the German press. What did they write about you?"

Goy handed him the copy of the *Fachzeitung Papier* he had brought with him.

"There you are. Has any decision been reached in my case?"

"As far as I know, the legal staff has come to a decision, yes. But before they make you a firm offer, they want to discuss the matter with you and sort out one or two details." László looked at his calendar. "The meeting could be held next Tuesday—could I suggest here in my office?"

Goy did his best to make conscientious preparations for the encounter. He had already submitted the contract signed in 1936, and he now gathered up all the supporting evidence: Bíró's letters and his replies to them, receipts, documents pertaining to the cooperation with the factory in Munich, and so on. As he stepped into László's office, however, he was unsettled by the sight of the veritable host of lawyers sitting around the negotiating table, representing the Ministry of Foreign Trade, the Patent Office, and various import-export companies.

The progress of Hungary's economic activities was what was generating all this interest. It was in this period that plans were laid for such large-scale projects as an atomic power plant and a major chemical manufacturing plant; a grand new coal-mining program had begun, and there was already mention of the dam scheme at Bős-Nagymaros. The Hungarian forint was very far from being convertible, and the implementation of these projects demanded more foreign currency than the Hungarian national economy was able to generate. The country's leadership was, therefore, on the lookout for any further sources of hard currency income. We can reasonably assume that it considered the court case surrounding the rights to the ballpoint pen to offer one such opportunity.

First and foremost, the lawyers present wished to clear up the background behind the contract signed with Bíró. Goy again gave his summary of events and handed the lawyers the supplementary documents, which they read on the spot. They were not, however, able to come to a clear decision—though it is unclear whether this was because of genuine caution or just the general reluctance to close a case for which the legal profession is so famous. It was the delegate of the ministry, Dr. Gábor Ürmösi, who summarized their opinion.

"We believe that because of its international connotations, this material also needs to be seen by a foreign lawyer. If Faber-Castell is in any case involved in the progress of this matter, we would like to ask them to have these documents examined by their own legal advisor or a lawyer they appoint. If he is in agreement that we have an adequate basis to set litigation in motion, then further decisions and expenses will no longer be their responsibility, but that of the Hungarian government."

It was Géza László who interjected with a question.

"And who would be the plaintiff in that instance?"

"These are flights of fancy for the moment, Géza, but I would not rule out appointing you for this task—not as a private individual, of course, but as the director of a patent company controlled by the state."

Dr. Zelen, Goy's former legal advisor, was sitting at the end of the table. He was not in agreement with this suggestion.

"In my opinion, Faber-Castell's original strategy was much more promising. Mr. Goy should institute the proceedings from Nuremberg. With a legitimate passport and a German work permit he can act as a fully empowered plaintiff."

Though no one said it out loud, all of them thought the same thing. If this was the path they took, they would have to accept responsibility were Goy, having traveled to Nuremberg with his family, to ask for political asylum and continue the case in his own name, for his private benefit. As a means of procrastinating the real decision, they decided that Goy was to go to Nuremberg in the company of Dr. Ürmösi to try to gain more information from Faber-Castell.

Sattmann welcomed his guests in the director's office. Ürmösi explained to him that were the case to come to trial, they would only require the advice of a German legal expert, and the expenses and risk would be covered by the Hungarian state. Sattmann was somewhat taken aback.

"I, for one, am only too glad to hear that the Hungarian state has adequate resources to decline our offer of financial support. I am certain that the Federal Republic of Germany would not be in a position to do the same thing. But I respect your decision. The interests of German pen manufacturers are affected by the case, however, and we will provide all the help we can give. If I may, I would recommend the services of our legal advisor, a man by the name of Dr. Gustav Maas. If you would like, I can ask him whether he can see you now."

Sattman put through the call. The lawyer was at his desk, and was willing to see the Hungarians.

Maas's office occupied the entire second floor of an old three-storey building, whose lobby led to the director's offices, the company archives, and the meeting room. The lawyer's office was decorated in the most spartan style. There was not a single item in it that might have served to distract one's attention from the subject of discussion; the only decoration on the wall was the West German flag. The gray-haired but youthful Maas was half done eating the slices of a peeled apple. On Goy and Ürmösi's arrival, he had his secretary remove the tray and immediately got down to business.

"Can I see the documents, please?"

He flicked through the dossier, taking a quick glance at every page in it, until finally he lifted out the German translation of Goy's contract with Bíró. He read it a number of times before speaking.

"That's the gist of it—the rest is just dressing," he said, putting the document to one side. "Herr Sattmann tells me that you want me to represent you in a case against the Biropatente Company. Is that right?"

"Yes, that's why we're here."

"I don't imagine you're expecting my decision on the double. I will look through this material carefully, and will inform you in writing of my decision within two weeks."

With this, Maas clearly considered the meeting to be over, but Ürmösi was insistent.

"I don't want to be pushy, but it would be very useful for us to know what, at first glance, you think our chances of winning the case are."

Maas shrugged his shoulders.

"My dear sir, that is a question for a fortune teller, not a lawyer. We can be as convinced as we like that we have a sure thing on our hands, but we are only victorious when we have convinced the court of the same. We will only know for certain when the last appeal goes in our favor."

"But could you estimate our chances?"

"This case promises to be a harder nut to crack than most. I am aware of the position held by the Biropatente Company. They will fight to their last man to defend their profits, and they have enough money to get the very best professionals on their side. As it happens, I have a question: who would the plaintiff be? Mr. Goy, I presume?"

"We have not yet made a final decision in this matter."

"It is not for me to interfere in your affairs, but for my part I think it crucial that Mr. Goy face up to Biropatente himself."

Maas stood up from his chair.

"Well, gentlemen, it has been a pleasure! My secretary will give you the warrant of attorney to fill out. If I decide to take your case, please return two copies of the warrant signed with an official stamp."

"You would ask for an advance, I imagine."

"It is too early to talk about money. When the case has been brought, I will send you the bill."

Back in Hungary, they continued to try to gauge their chances of success. The lawyers picked through the various eventualities over and over again. They determined that Biropatente would have a list of counterarguments up its sleeve. The wording of the contract only referred to the "so-called Biro pen"; they would argue that this was not one and the same as the ballpoint pen they had since patented all over the world. From Goy's point of view, it was a mitigating circumstance that two of the three fundamental principles behind the working ballpoint pen were already present in the early Hungarian version. It was also a favorable consideration that he had invested a significant amount of money in the patent application at the time, and that he had himself participated in the technical improvement of the pen.

The fact that Goy had taken twenty years to defend his rights would work strongly against him, however. The question of statutory limitation would inevitably arise and the relevant civil laws would have to be examined in each country in question.

The final conclusion they reached was that while the basic contract could not be considered ineffective, there was not a strong enough case to maintain Goy's share of the proceeds at the original fifty-fifty level.

The arising uncertainty was only dissipated by the letter Maas sent in the meantime. He announced that on the basis of the material he had examined he was willing to take on the case. The Ministry of Foreign Trade thus decided that proceedings should be begun. Rejecting all recommendations to the contrary, it nominated the Licencia Corporation as the plaintiff; therefore it was effectively the Hungarian state that was entering the fray. Goy had to cede all his rights to the Licencia Corporation. It was only in the division of the proceeds that he had a say—finally talking them into seventy percent for himself and thirty percent for the state. He only succeeded in coming to such a favorable arrangement because, under the regulations of the time, he would have received most of his due in forints, and the authorities were primarily interested in the level of foreign currency income from the proceedings. Maas filed the lawsuit. As the costs of the trial were proportional to the size of the claim involved, he diminished the risk by mentioning only a partial figure of ten thousand marks for himself for the time being.

CHAPTER 29

A t 4 Zrínyi Street, on the northern outskirts of Budapest, the Hungarian Pen Manufacturers' Cooperative was having the final touches put to its new two-storey building. The automatic plate machine and its accompanying apparatus were delivered by the Schneider company, and the dye paste that arrived from elsewhere exceeded all expectations. Modern Hungarian ballpoint pen production began in earnest.

The investment was enormously expensive by the standards of the time, costing one million convertible Hungarian forints—a fortune. Káldori wished to recover this money as soon as possible. The pens cost only eighteen forints, less than a good-quality imported pencil, and so they sold well in Hungarian shops, but Káldori wanted to break into foreign markets, too. He wanted to prove himself to those at the Ministry of Foreign Trade by generating hard currency income.

No sooner had the ink dried on the contract with Goy, which essentially acknowledged his rights to the pen, when the company director wrote a letter to the patent agent Miksa Aknai, the Budapest representative of Henry George Martin, Bíró's legal successor. Káldori asked him to grant them permission to produce the pen, and to name his price for the royalties. Goy was dumbfounded at this news.

"What need is there for that? We have just submitted our documents to the court in Nuremberg, in which we state that the license to produce in Hungary is mine."

"This is true, but for the time being Martin still has the patent rights here, and unless we get his permission, we cannot export anywhere, and that is our chief goal. It's no good getting the goods out through other channels, for they can always get the police to confiscate them. But if you really want to know, this isn't my idea—the ministry asked me to get in touch with them."

"I could only support this undertaking if we were talking about one of the countries that was assigned to Bíró in the original contract, like those of South America, but this isn't true of the European countries you are interested in."

"I honestly hope that the court in Nuremberg will rule in our favor as soon as possible, and that the whole problem will take care of itself. Until then we have no choice but to abide by the regulations in force at the moment."

When the reply arrived from Martin's agent three weeks later, Goy could not help indulging in a little schadenfreude at the extraordinarily high royalties demanded by Aknai. His offer was ten percent of the sales price of the cartridges, irrespective of whether they were complete pens or just replacement cartridges. A further condition was that the words "Bíró license" be inscribed on every pen.

Káldori stuck to his guns. In the hope of coming to a more favorable agreement, he asked to hold a private meeting with Dr. Weiss, a lawyer from Zürich, one of the directors of the Biropatente Company. They agreed to meet in Nuremberg in late October 1959. In the end the rendezvous was cancelled: Káldori was overcome with doubts as to what his strategy should be. He sent a message to Weiss, saying that he had urgent business in Vienna—could he look him up there? The lawyer did not accept this invitation, saying he had other engagements, but offered to keep the line of communication open, suggesting a meeting for November 30, this time in Switzerland. Káldori rejected this outright. He was no doubt encouraged in this decision by the news that the court in Nuremberg was beginning to deal with the case the Licencia Corporation had submitted.

Goy was called as a witness to the trial; he knew Nuremberg almost like the back of his hand by now, and booked a room at his usual lodging, the Hotel Post. He had hardly unpacked his things when a call was put through to his room. It was the owner of a Nuremberg metal factory by the name of Meier, who wanted to meet him urgently. Goy did not know him personally, but did not reject his offer out of hand, thinking that he might be able to provide valuable information. He asked for an hour to think the matter over, during which he made a call to ask for Maas's advice. The lawyer had not heard of Meier, but saw no particular risk in getting to know him.

The metal works were on the city outskirts, and Meier sent a driver to collect his guest. He was unable to welcome Goy immediately, as he was in a meeting with a regular customer of his, and entrusted his nephew with showing the Hungarian around the plant.

The facility housed six old-fashioned drawing tables powered by transmission belts, with one worker behind each, inserting copper pipes into it. Collapsible ballpoint pen casings were produced here for traders at fairs. Meier, who had finished his meeting, ran to greet Goy at the end of the building.

"I don't imagine you've seen too much of interest here. We get what we can out of these poor old wrecks for as long as they'll bear it, then they're off to the scrap-heap. Let's go up to my office for some decent coffee and a cognac."

While they were sipping their drinks, Meier glanced up regularly at the wall clock. Goy could not stop himself from quizzing him.

"Are we waiting for someone?"

"Yes, a gentleman who could be of great importance to you."

"That all sounds rather secretive to a simple man like me. Could I ask who it is that wants to meet me?"

"I read the articles about all this in the German Press, and as someone in the profession, I was interested in meeting the man who pulled László Bíró's chestnuts out of the fire."

Not satisfied by this explanation, Goy shrunk back, giving only monosyllabic responses to Meier's further questions. He was just thinking how he could possibly make an excuse to leave when the door opened, and an elegant man of around seventy entered the room, wearing a bow tie and spectacles on a string.

"Mr. Goy?" he asked, hardly raising his voice, clearly wishing them to pay close attention to what he had to say.

"Yes. To whom do I owe the pleasure?"

"I am Dr. Hugo Weiss, from the Biropatente Company."

Goy did his best to hide his shock at finding himself face to face with the leading member of his legal opposition.

"So it was you who engineered this meeting. What was all this conspiring good for?"

"If I had invited you myself, would you have accepted?"

"Probably not."

"You see." He gestured to Meier. "You can go, Edward. Send in a blümchen coffee, would you."

"What kind of coffee is that?"

"It's so weak you can see the flowers at the bottom of the cup."

Once they were alone in the room, Weiss let his gaze wander all over the expression on Goy's face.

"So you're the *blöde* Hungarian?" he asked, his voice as reserved as before.

"If you'll pardon my asking, why am I mad?"

"Because you insist on hitting your head against the wall, even though we all know that walls are stronger than heads."

"Actually, my experience has been that if you really put your mind to it, the wall will often fall back in front of your eyes."

"On what do you base this hypothesis?"

"For example, on the fact that you are here."

Weiss chuckled.

"That's good, very good."

"And if you don't think it rude of me, could I ask you to speak a little more loudly?"

"So you want you take away every tactical advantage at an old man's disposal? As you wish. It would have been better if we'd met earlier, but maybe it is not too late for us to come to see eye to eye." Weiss took out his notebook. "What else is there? You have played your cards well—you have employed a number of tricks, getting a few German pen producers on your side, mobilizing the press, but I can assure you that you will not win the hand. Reynolds is a big company, but we succeeded in exacting the two million dollars in royalties they owed us."

"If you are so sure of your success, what is it that you want from me?"

"There is no such thing as one hundred percent success in a legal case. We do have something to lose; the longer it lasts, the greater this loss will be. The trial itself will inflict serious financial damage on the Biropatente Company, and the news of the scandal may also spur other clients who are unhappy with our conditions to take action. This is why I would ask you to try to come to some agreement."

"You know very well that this is not my case but that of the Licencia Corporation, or rather, through them, the Hungarian government."

"I would be honored if you would not regard me as an amateur. The whole of the plaintiff's case is built around you. If you were to step out of it, it would fall apart of its own accord. But we do not wish you to do anything that might risk your livelihood. There is no need for an obvious about-face—it is enough if you make one or two 'accidental' slip-ups while giving evidence."

"The members of the Hungarian leadership are no fools. They won't just leave it at that."

Weiss smiled.

"You think so? Well, suffice to say you are not the first person in the Hungarian team we have talked to, and so far we have found common ground with everyone, in half the time the two of us have been speaking now."

"I don't believe that."

"This is a free country, you know—you can believe whatever you like." Weiss again peered at his notebook. "What else is there? Ah, yes. It is conceivable that you are partly motivated by personal considerations; you might want to exact your revenge on Mr. Bíró, I mean. I would agree with you that as an individual Bíró did not treat you fairly. He should at least have given you back the few thousand dollars that you gave him in cash in the course of his research. But this emotional factor has no foundation, for our poor ill-starred Bíró is a part of the company by name alone. In practice he has nothing to do with it, for he has no shares. We shouldn't make too much of this, of course, as his name still carries serious weight."

Instead of deriving satisfaction from this, Goy instinctively felt sympathy for his former colleague. He was sad to think that every trace of their joint efforts had been lost, and that the fate of the ballpoint pen had forever fallen into the hands of strangers.

"Now that we've got that over with," Weiss continued, "let's finally talk numbers. Your current claim is for ten thousand marks. According to your contract you would receive seventy percent of that—or, rather, a little less than that, because you would be paid in forints, and so lose on the exchange rate."

"I see you've done your homework."

"Thank you. We would offer you, say, thirty times this amount, two hundred thousand marks. We would deposit the money with the solicitor's office of your choice and we would even be willing to give you an advance on it. On the day the judgment is declared, the complete amount would be there for you to pick up. You could collect it in cash or, if you prefer, mandate a *treuhand,* a middle-man. What else is there? We do not need any kind of receipt or invoice, as discretion is in our interests, too. And so there is no risk of getting caught. I would be only too happy to give you time to reflect on the matter, but the circumstances make that impossible, I'm afraid. I await your response in earnest."

Goy would have liked to drink another cognac, but was afraid Weiss might see this as a signal of assent, a toast. He turned away and thought the whole matter through. This contract with the ministry would give him more money than he was being offered now, but that money, if ever he could get his hands on it, was hidden in an uncertain future. Back in Hungary he continued to earn 1900 forints a month, and was hardly able to cover his family's most basic needs. The advance being offered by Weiss, alone, would cover his current financial worries while, in the longer term, it would pay for his children to be educated at good schools abroad. And yet, as if directly from the pit of his belly, he felt an instinct of objection rising up within him: the old man felt that were he to agree to this bargain, he would be tortured by self-recrimination for the rest of his days for not having fought the good fight to the bitter end. He shook his head, and in a voice so pained that it seemed alien even to himself, he made his reply.

"I cannot accept your offer, Mr. Weiss."

"I am not used to haggling, but we could make a modest improvement to the conditions I mentioned."

"That is not the issue. I want to win."

"Like David against Goliath?"

"Something like that."

"I think the press at the time falsified the outcome of that particular contest; there was no way David could have won. You will see that for yourself. Thank you for your time. I enjoyed every moment of our little conversation."

CHAPTER 30

The attempt at a deal unsuccessful, a lengthy court case commenced, with the recesses between the various court sessions often lasting for four or five months. It was not in the interests of Biropatente to speed up the pace of events. It stood its ground and continued to enjoy its income, hoping to protract the verdict until 1966. In that instance it would have nothing to lose, as that was when the patent protection for the ballpoint pen in Germany would expire of its own accord.

To begin with, Maas and his colleagues tried to unnerve their opponents by submitting the documents pertaining to earlier cases brought by German and French manufacturers against Bíró or the companies in his hands. Some of these had succeeded in their legal battles, like Walter Langweiler, producer of the Floball pen, who had not had to pay any royalties to Biropatente ever since.

Goy, the plaintiff's chief witness, was cross-examined on a number of occasions; he was asked to recount every detail of his relationship with Bíró. His testimony essentially repeated the information included in their submission to the court, but the third time he produced a new document. A few days earlier, a letter from 1939 had been found in the archive of a patent agent in Budapest, one in which Bíró offered his ballpoint pen invention to a country where 50% of the distribution rights belonged to Goy. The ageing witness argued that this document, casting doubt as it did on the inventor's professional honor, could also provide an explanation for why Bíró did not wish to contact him.

"It is obvious that he would have had trouble explaining this behavior," said Goy.

The team of experienced lawyers on the other side, led by Weiss, decided that attack was the best form of defense. They sniffed out what they saw to be the plaintiff's Achilles heel: its right to represent itself. They argued that the Licencia Corporation was not in a position to represent Goy's interests, for moral reasons, if nothing else, because it was in the hands of a state that nationalized his factory in 1949 and his private house in 1952. Biropatente's lawyers claimed that there was no guarantee that, should they win the case, the Licencia Corporation would not try to rob Goy of the rights granted him by his contract with Bíró. The court gave considerable weight to this argument. Maas would afterward repeatedly pour scorn on his Hungarian clients.

"This would all have been avoided had you accepted my advice and nominated Mr. Goy as the plaintiff. We would have had it all stitched up ages ago."

"It wasn't my decision, believe me!" Goy replied to the criticism.

"It doesn't make any difference now. Our only task remains to fight for justice to the bitter end."

Weiss and his team were not satisfied with this first advance; they wanted to seal their victory. The ammunition turned out to be closer at hand than they had expected when they found the letters that had been exchanged with Hungarian companies in their files. They first presented the 1959 application for a license from Endre Káldori, president of the Hungarian Pen Manufacturers' Cooperative, and then a similar request from Chemolimpex, taking over the representation of the cooperative; the latter had been submitted while the trial was in progress. The Swiss side argued that with this, the Hungarians had acknowledged the Biropatente Company's rightful ownership of the license. They emphasized that Goy was himself a member of the cooperative in question.

There was nothing Maas could do to dodge these blows. Instead, he tried to save face by bringing into question the originality of Bíró's whole invention, listing earlier contributions for all the elements of the pen. But in his heart of hearts he had given up the struggle. In a recess during one of the hearings he sat down to talk to Goy and the representative of the Licencia Corporation.

"A doctor has the right not to tell his patient the truth about his condition—sometimes this is how he can help the most. But a lawyer must not start building castles in the sky. I have to be straight with you. I don't want to disappoint you, but this case is as good as lost. All I can do is repeat for the hundredth time what I said to begin with: we could not make up for the handicap arising from the wrong choice of plaintiff, and the superiority brought by Biropatente's financial clout."

"We can still win on appeal."

"In that event it will be necessary to inform the parish priest that a miracle has taken place. The court of appeal almost never turns over a decision of the trial court."

Goy stared into space in silence, while the man from the Licencia Corporation had no option but to accept defeat.

"In my opinion," said the Hungarian official, "you played your part in this loss. I regularly listened to you at the hearings—I never felt real passion in your voice."

"What you are thinking of is the theater or the circus, not a court-house; there is no place for emotions here. My conscience is clear. I did everything in my power in the given circumstances."

From then on events progressed quickly. Maas's prediction proved to be correct: the court rejected the Licencia Corporation's lawsuit both at the first round and on appeal.

The day after the final verdict was announced, Goy was just making his preparations to leave, when the hotel porter handed him a gift-wrapped box.

"This was delivered to you."

Goy opened the package. In it was a ballpoint pen, the casing of which was a replica of Michelangelo's David—it was part of the well-known series based on various works of art. A short note had been inscribed on the accompanying card:

"With admiration, Goliath."

Goy looked for the nearest rubbish bin and threw the pen straight inside.

Goy was confident that the unfavorable decision of the Nuremberg court would not affect his earlier agreement with the Pen Manufacturers' Cooperative. It did not, after all, set any conditions about the results of the trial. He thought he would receive five percent of the factory price of the ballpoint pens manufactured in Hungary.

He was to be disappointed, however. The cooperative refused to accept his claim. Káldori cited a clause whose purpose had previously not been clear—under which the contract would be eligible only if the Center for Craftsmen's Cooperatives gave it its approval. This supervisory authority was, it seemed, unwilling to give any sign either of agreement or rejection.

Goy tried to fight for what was rightly his, requesting a legal opinion from Dr. Ferenc Windisch, the former head of the legislative department at the Ministry of Justice. Windisch's professional evaluation stated that the laws in force at the time the agreement came about between the Pen Manufacturers' Cooperative and Goy did not demand the approval of a regulatory body. As both parties began and continued to follow the stipulations of the contract even when it was obvious to them both that the approval of the overseeing body would never arrive, their rights and obligations under the agreement were still binding.

When the cooperative's leaders maintained their stance despite this professional opinion, Windisch advised Goy to take legal action against them to defend his interests. The old man would have had trouble taking this step; he did not want to be unfair to those who had provided his livelihood for so many years. His decision was also influenced by the news that Káldori was suffering from terminal cancer. He made a further attempt at conciliation, writing a letter to the cooperative in which he described how the benefits of the oft-mentioned Bíró contract were enjoyed by Hungary as a whole and by the cooperative in particular, for which it brought significant savings. The Biropatente received or extracted royalties in every country in the world where ballpoint pen production took place, and Hungary was the only exception to this—precisely thanks to the contract in question. In return, Goy thought it equitable to receive an amount of money equivalent to five percent of the cooperative's returns between January 15, 1959, when he had joined it, and the end of 1969, and to be given a premium of two thousand forints a month from 1970 onward.

Goy's submission was discussed at the meeting of the leaders of the cooperative. They decided that it was not in their power to grant his request, but out of respect they awarded him a one-off sum of twenty thousand forints "in recognition of his services to the field of ballpoint pen production."

By this time Andor Goy had reached the age of seventy—he did not have the strength to fight on. He collected the money, and with that he regarded the case as closed. He remained as an employee of the cooperative, but soon took leave to accept a job offer from the Schneider company in West Germany. They asked him to serve as a consultant in the development of a new pen.

The idea of staying in Germany permanently did cross Goy's mind. He spent a number of weeks there, but despite the relative luxury of the surroundings he somehow did not find his feet. Hard as the Germans tried to persuade Goy to stay, his thoughts continue to revolve around Hungary.

Chief executive Christian Schneider took him out to the station, and as they still had time before Goy's train left, he invited him to the snack bar for a last cognac for the road.

"To your good health, my friend. *Auf Wiedersehen!*"

"Thank you for everything, Herr Schneider, but I don't think I could manage another trip like this. I've had enough, I've really had enough."

"You mustn't lose heart. God is not a construction manager, who settles accounts with his workers every Friday, but in the long run he always sees to it that justice is done."

"There was a time when I also used to think that there was order in the world, but since then how many times have I seen manipulation and intrigue emerge victorious and genuine achievement go unrewarded? Let's not beat around the bush: the ballpoint pen has made hundreds of thousands of people rich, and left me without a penny."

Schneider had nothing to say to this, and could only shrug his shoulders. The ensuing silence was broken by the sound of the megaphone announcing the arrival of the express train to Vienna. Goy stood up.

"*Zeit abzugehen!* Time to go!"

CHAPTER 31

I f Elza had not done such a good job in saving for a rainy day when the family was enjoying better times, they would have been in despair after the loss of their Birome shares. As it was, László Bíró was able to pursue the development of new ideas in reasonable financial security.

It was during this time that Bíró made a thorough examination of the plastics which made up the ballpoint pen casing. He used what he had learned to produce a significantly different version—one that surpassed all former models, even bakelite ones, in its shine, strength, and resistance to solvents. He called his new invention the Birolit. Bíró invested the income it was to generate in new research. He developed a new kind of double-filtered holder for cigarettes, which removed seventy-four to seventy-eight percent of the smoke's damaging elements. As a sign of respect, he sent the first of these health-conscious devices to none other than the president of Argentina, Juan Domingo Peron. He had enjoyed good relations with Peron from the time when he was minister of industry. Peron would often seek Bíró's opinion, and in return the inventor could count on his help with any of his problems. It was not long before Peron invited Bíró to pay him a visit.

"I'd like to thank you for this filter. It is Evita who is most grateful—she doesn't have to hear me hack away every morning."

"Yes, it was my wife, too, who made me work on this. I kept promising her that by my sixtieth birthday I'd only smoke sixty a day—you know what women are like!"

"How right you are. But this isn't why I invited you here. As far as

I'm concerned, the name László Bíró is already synonymous with the successful and productive representation of Argentina, by spreading the country's name with the ballpoint pen all over the world. Now, however, I would like to ask you to fulfill a more formal diplomatic mission, for one time only. When where you last in Hungary?"

"Back in 1938, when I had to flee the country—that's almost thirty years ago. Since then the furthest I've got is Austria. Back then they regarded me as an 'undesirable element,' and I have never managed to forget that."

"I would like you to make another visit to your old homeland. A commercial-industrial delegation is going to Budapest, and I would like you to lead it. As you can imagine, I didn't want to trouble you with this, but I simply can't find anyone suitable apart from you."

"Mr. President, back then, I made a promise never to step back across the Hungarian border."

"There have been bigger promises broken since then." Peron smiled. "If it would reassure your conscience, I could speak to Isidro Klein, the chief rabbi of Buenos Aires, and ask him to absolve you of your pledge."

"I know him well enough to know he will say that I swore not to step across the border, and crossing by plane does not apply."

They both laughed. Peron extended his hand.

"Think it over. I would be very glad if tomorrow your answer is yes. You would be accompanied by your wife, of course."

Bíró expected Elza to join him in rejecting Peron's offer. To his surprise, it was she who convinced him to accept.

"We should be glad for the chance to make one last trip. We are not poor, but we could not afford such a journey out of our own money."

Bíró made no reply. Elza turned his face around toward her.

"Who are you angry with? More than half of the people in Hungary were not even alive at the time we had to leave. It's about time we started making peace with them."

"Is there no other reason why you want to go to Budapest?"

His wife looked down, bashfully.

"Actually, there is. Klári, my cousin who used to live in Hungary and now lives in Australia, is visiting Budapest now. I would like to meet her."

The meetings in Budapest went on from one day to the next without producing anything in the way of significant results. The enormous distance between the two countries effectively ruled out large-scale economic cooperation. Bíró soon became bored of this mock activity, and increasingly entrusted the handling of the talks to his subordinates.

The inventor strove to remain incognito while in his home city, but one Hungarian journalist did track him down to his hotel room. Bíró happened to be suffering with the flu, and did not have the strength to wriggle out of the interview.

"Have you ever been honored for your work?" the journalist asked.

Bíró stared blankly.

"I was paid my due. What do you mean?"

"Any kind of award. Here in Hungary, for example, they would give someone a Kossuth Prize for an invention as significant as the ballpoint pen."

"Oh, I see. No, I haven't received anything like that."

"You would deserve it, though."

"Deserve it? Why?"

"In recognition of your achievements, your successes."

Bíró shrugged his shoulders.

"Only fools are interested in successes. Anyone who has taken a creative process from start to finish will know from first hand how little it would have taken for it to have failed along the way."

"What do you think is a real success, then?"

"The fact that you are holding a ballpoint pen in your hand as we speak. It means that millions of people can get hold of a pen cheaply and easily. Just as much a boon for democracy as the transistor radio or the quartz clock. The bastions of privilege based on money have fallen."

Once recovered from his flu, Bíró would often go out for a walk with his wife. He found that everything in the city had changed; he could only find the place where their old apartment had been. The two-story building had been replaced by a drab, concrete housing project, and the Red Salon of the Hotel Royal, where they had once gone for their first dates, had long been closed.

"The only things that haven't changed are the sky and the Danube," said Elza.

One day Bíró asked that they pay a visit to his father's grave in Budapest's largest Jewish cemetery. He bought a bunch of flowers from one of the stands near the entrance, but looked so awkward the way he held them that Elza took them from him. He handed a few dollars to the caretaker, who willingly searched through the records for the lot the elder Bíró was buried in, and even drew them an impromptu map of how to get there.

On the way they stopped at a marble plaque. Bíró read the inscription: it had been erected in memory of a group of Jewish forced laborers who had been executed by the fascist Arrow Cross party in the last days of the war.

"They were the same age as me. If I hadn't fled abroad . . ."

"Don't say things like that. Let's go on."

Mátyás Bíró was buried in a corner of the cemetery that had long been filled with graves. Jewish cemeteries do not observe the common European tradition of reusing burial places after some years or decades, and the elder Bíró's bones had been left intact. But, left unattended over so many years, the entire lot had been overgrown with brush and clematis; Bíró asked a nearby gardener to cut him a path to his father's grave. The marble slowly began to peep out of the undergrowth. In the course of one harsh winter it had cracked, whereupon the gold paint had disappeared from the letters and the numbers, rendering them unreadable.

"Don't forget to remind me that we should give some money to the synagogue for them to do it up."

"I won't forget, Laci."

Bíró, true to Jewish custom, put a stone on the tombstone in remembrance, before gesturing to his wife.

"Let's sit down a while."

He closed his eyes and turned his face toward the early autumn sun. His temple was sweating under the brim of his cap. Elza gave him a handkerchief.

"The sun is so strong! The weather is always good around the time of Jewish festivals."

Bíró struggled to find words.

"There's something I'd like to tell you. You were in a sanatorium bed at the time my father was dying. He had been unconscious for three weeks, in a coma, and my brother and I took turns at his bedside. One night I saw that he had regained consciousness, but it was obvious from his expression that he was in great pain. I called for the doctor on duty, who diagnosed cardiac arrest, and asked for adrenalin, which he wanted to inject into his heart with a long needle. I grasped his hand and asked. Would this injection keep my father alive? No, he said, at best it would give us a bit of time, enough for the family to arrive and pay their respects. 'Then don't give it to him, doctor,' I said. Within a minute my father was dead. Since then I have often wondered whether this was the right thing to do, and I think it was. By then he had already been through the moments just before death, which are the most painful ones. What right did I have to make him live through them all over again?"

"Why are you telling me this now?"

"We have to talk about this sooner or later, Elza. If I come to that same moment, please allow me to die. You have been my partner through everything, you've seen what a good life I've had. I have been able to exert my every ounce of strength and energy usefully. When finished, my body will be laid down, empty."

Elza began to cry.

"You promised me you would never leave me on my own!"

"We did not arrive together, and neither will we leave together. But let's not mention it ever again."

He helped his wife up from the bench.

"Come along, my love."

EPILOGUE

The older Bíró became, the more he worked. He worked for the largest Argentinean producer of ballpoint pens, Sylvapen, as an advisor and advertising director. It did his ego no harm to face a life-size statue of himself in the company's foyer every morning. In addition, as a member of the Argentinean Atomic Energy Committee, he experimented with new technologies for the refinement of uranium—it was with this that the country sought to defend itself against Arab oil threats.

One day Bíró became unwell at home. His heart began to beat irregularly and he lost consciousness. He was transported to the German hospital in Buenos Aires, where he was given a tranquillizer injection, but this could not help him. He died on November 24, 1985, at the age of eighty-five. Not long before his death he said to his son-in-law, a teacher by profession:

"The notion of death is important for us to live better. This should be talked about in schools, too. This would remind people what a short time we are given to live, and give them a perspective that would lead to them rethink concepts like democracy, social justice, and humility."

Argentina was to keep Bíró's memory alive. According to the official tributes paid to him, he was without doubt the country's most significant inventor, who in his public role had made a commitment to the furthering of human progress. A few years later, his birthday, September 29, was made a public holiday dedicated to Argentina's greatest inventors.

It was the older of our two protagonists, Andor Goy, who was to live the longest. For many years he would travel regularly to the northern outskirts of Budapest, where he worked as a machine designer at the ICO manufacturing facility. Some remember him as having often personally operated certain equipment. Later he translated short technical documents from German for the Center for Self-Employed Craftsmen, and he guided foreign delegations around the country. Even at home he was incapable of taking his mind off work, drawing lines with a ballpoint pen on pieces of paper while he sat in front of the television, thus testing the capacity of the ink cartridge. He was in his ninetieth year when the Hungarian government awarded him the title "Distinguished Worker."

Goy was all too often beset with financial woes. On one occasion he even had to travel to Vienna to sell his cherished set of billiard balls. He thought they were made of ivory, but when he got to the shop it turned out they were plastic. The money he received in exchange was hardly enough to cover his lunch at the train station. He did, in general, enjoy his food and quite a lot of it at that. He was fond of strong, mature cheeses and spicy paprika, and would have fresh-baked bread with whatever he ate.

It was in 1991, on the name day of his daughter Gabriella, that Andor Goy fell ill. He went to the bathroom and collapsed, face down, by the tub, never to recover his senses. He was ninety-six.

The fiftieth anniversary of the first marketable, mass-produced ballpoint pen was celebrated in 1994 in Argentina on a grand scale. An international exhibition was held in the premises of the main post office in Buenos Aires. There were many sensational moments during the series of events, like the German Hans Georg Schniever displaying the world's largest collection of ballpoints, all thirty thousand of them, some of which were even edible.

Also present was Károly Misley, Hungarian ambassador to Argentina, who had formerly been a close friend of Bíró's. It was he who first had the idea of honoring the inventor's career back in Hungary. He convinced Mariann Bíró of his plan, persuading her to come to Budapest and help open the event, and to provide him access to her father's personal effects and the official documents still in her possession.

While the technical difficulties did not appear to be insurmountable, there was a moral obstacle of sorts: the uncertain response of the Goy family. Andor Goy did not forget the injustice he had suffered to his dying day, and his ancestors felt obliged to perpetuate his view, as if it were an inheritance. At the news of the planned Budapest exhibition, journalists paid a visit to the Goy family home, asking Ferdinanda, Goy's seventy-nine-year-old widow, for a statement.

"They cheated my husband out of everything!" she said. "The idea was Bíró's, and I wouldn't, and couldn't say a bad word about him, but it was my husband who developed it into a usable product. And that is why it hurts me so much that nowadays his name is never mentioned in association with the ballpoint pen."

The newspapers predicted a scandal. These are, of course, exactly the sort of prophecies that tend to be self-fulfilling. The opening of the exhibition, on June 24, 1996, at Budapest's Palace of Technical Sciences on Kossuth Square, was awaited with the utmost anticipation.

Goy's widow wanted to avoid even the semblance of provocation, and stayed away, leaving only her two daughters to represent the family. The young women retired to a corner of the hall, as if to symbolize the distance they intended to keep from the proceedings.

In the minutes before ten o'clock the guests of honor arrived in turn—the Argentinean ambassador to Budapest and his Hungarian counterpart in Buenos Aires, and many prominent figures from public life. Mariann Bíró had spent the night at the residence of the event's organizers, and had fallen asleep from jet lag. She arrived a bit late in the grand lobby the event was being held in. Holding a huge bunch of flowers in her hands, she rushed over to Misley and asked him something; the ambassador pointed her in the direction of the Goy girls. Mariann Bíró stopped next to Gabriella and Krisztina Goy, split the bunch in two, and handed it to them.

"Let's settle our differences," she said, "for the sake of the ones we loved."

And, at that, their movements tentative at first, the three women held each other in a tight embrace.

PRIMARY SOURCES*

Bíró, László József. *Csendes forradalom. A golyóstoll regénye* [*Silent Revolution. A Novel about the Ballpoint Pen*]. Hungarian trans. Tibor Kóródy and Lajos Pálfi. Európa, Budapest, 1975.

Braham, Randolph L. *The Politics of Genocide: The Holocaust in Hungary*. Columbia University Press, New York, 1981.

Előre (*Forward*), a Budapest weekly magazine. Issues from 1936 to 1938.

Feuerstein Emil. *Egy marék virág I-II* (*a magyar ajkú zsidóság szellemi öröksége*) [*A Handful of Flowers I-II* (*The Intellectual Heritage of the Hungarian-speaking Jewry*)]. Private publication, Tel-Aviv, 1986.

Garay, Tóth János and Nagy, Ferenc, eds.: *Holográfia és humanizmus. A Nobel-díjas Gábor Dénes* [*Holography and Humanism. Noble Laureate Dennis Gabor*].

Goy, Andor. *Így igaz* [*The True Story*], manuscript.

Kertész, Árpád. *A nyomtatott betű története és útja Magyarországon.* [*The History and Path of the Typed Letter in Hungary*]. Magyar Nyomdászok Társasága (Association of Hungarian Printers), Budapest, 1941.

Kurucz, Ladislao (László Kurucz). *Los húngaros en la Argentina–A magyarok Argentínában.* [*Hungarians in Argentina*]. Bilingual Spanish-Hungarian edition. Ediciones Biblioteca Nacional, Buenos Aires, 1999.

*Among the author's numerous other sources were newspaper articles, letters of patent registration, and interviews held in Hungary and Argentina.

Nagy, Ferenc, ed. *Neumann János és a "magyar titok"* [*John von Neumann and the "Hungarian Secret"*]. Országos Műszaki Információs Központ és Könyvtár (National Technical Information Center and Library), Budapest, 1987.

Nagy, Ferenc, ed. *Harsányi János, a játékelmélet Nobel-díjasa* [*John C. Harsányi, the Nobel Laureate of Game Theory*]. Akadémia, Budapest, 1995.

Nagy, Ferenc, ed. *A Nobel-díjas Békesy György* [*Georg von Békésy, Nobel Laureate*]. Better, Budapest, 1999.

Szabó, Zoltán. *Szellemi honvédelem* [*Intellectual Defense of the Nation*]. Héttorony, Budapest, 1994.

Szentes, Éva and Hargittay, Emil. *Irodalmi kávéházak Pesten és Budán* [*Literary Cafés in Pest and Buda*]. Universitas, Budapest, 1998.

Vámos, Éva. *László Bíró*. Novofer Foundation, Budapest, 1995.

Vámos, Éva. *"Kiállitás Bíró László József magyar-argentin feltaláló életművéről, a golyóstoll történetéről"* [*Exhibition on the Life Work of the Hungarian Argentine Inventor László József Bíró and the Story of the Ballpoint Pen*]. Article in *Technikatörténeti szemle* (*Review of the History of Technology*), vol. 22, Budapest, 1996, pp. 281–285

INDEX